ENERGY EFFICIENCY IN INDUSTRY

Proceedings of a workshop organised by the Commission of the European Communities, Directorate-General for Energy, held in Berlin on 19th and 20th October 1987.

ACKNOWLEDGEMENT

Particular thanks are due to Mr G. Vacchelli, consultant to the Commission of the European Communities, for editorial assistance concerning the discussion.

ENERGY EFFICIENCY IN INDUSTRY

Edited by

J. SIRCHIS

Directorate-General for Energy,
Commission of the European Communities, Brussels, Belgium

ELSEVIER APPLIED SCIENCE
LONDON and NEW YORK

ELSEVIER APPLIED SCIENCE PUBLISHERS LTD
Crown House, Linton Road, Barking, Essex IG11 8JU, England

Sole Distributor in the USA and Canada
ELSEVIER SCIENCE PUBLISHING CO., INC.
52 Vanderbilt Avenue, New York, NY 10017, USA

WITH 17 TABLES AND 98 ILLUSTRATIONS

British Library Cataloguing in Publication Data

Energy efficiency in industry.
1. European Community countries.
Industries. Energy Conservation
I. Sirchis, J.
658.2′6

Library of Congress CIP data

Energy efficiency in industry.

(EUR; 11 490)
Text in English; summaries in French and German.
"Proceedings of a workshop organized by the Commission of the European Communities, Directorate-General for Energy, held in Berlin on 19th and 20th October 1987"—P.
Bibliography: p.
Includes index.
1. Industry—Energy conservation—Congresses.
I. Sirchis, J. II. Commission of the European Communities. Directorate-General for Energy.
III. Series.
TJ163.27.E515 1988 621.042 88-16044

ISBN 1-85166-243-X

Publication arrangements by Commission of the European Communities, Directorate-General Telecommunications, Information Industries and Innovation, Scientific and Technical Communications Service, Luxembourg

EUR 11490

Printed in Great Britain by Galliard (Printers) Ltd, Great Yarmouth

PREFACE

The competitive pressures on industry have never been greater and no company can afford to ignore ways of reducing costs or of using its resources more efficiently. Energy is an input of major significance for many industries, and an appreciable cost element for many others. Furthermore, energy prices are a volatile factor which could again increase significantly in the future.

This conference, organised by the Commission of the European Communities, aimed to make firms fully aware of the different opportunities for improving the efficiency of energy use, and reviewed the latest techniques and systems including:

—process integration,
—industrial plant—process control and optimisation,
—new techniques for low temperature and heat recovery,
—the energy management of utilities,
—sources of finance for energy efficiency investments.

This volume contains the oral papers presented at the conference and the round-table-discussions.

CONTENTS

Preface v

OPENING SESSION

Opening address 2
G. Turner, Senator for Science and Research, Berlin

Opening address 5
G. Briganti, ENEA, Rome, Italy

Opening address 8
C. S. Maniatopoulos, Director-General for Energy, Commission of the European Communities, Brussels, Belgium

SESSION I: OVERVIEW

Ways and techniques in the rational use of energy 14
H. Schaefer, Institut für Energiewirtschaft und Kraftwerkstechnik, Munich, Federal Republic of Germany

SESSION II: PROCESS INTEGRATION

Energy savings in the manufacture of crankshafts—an example of integrated analysis based on detailed measurements 28
M. Rudolph, Professor of Power Production and Power-station Technology, Munich, Federal Republic of Germany

Process integration using pinch technology 37
B. Linnhoff, Centre for Process Integration, UMIST, and A. Eastwood, Linnhoff March Ltd, Manchester, United Kingdom

Process integration in a benzole refinery 52
R. L. Bardsley, Staveley Chemicals, Chesterfield, United Kingdom

The results of a process integration study to improve energy efficiency at a British brewery 56
R. Marsh, Chief Engineer and Energy Manager, Tetley Walker Limited, Warrington, United Kingdom

SESSION III: NEW TECHNIQUES FOR LOW-TEMPERATURE HEAT RECOVERY

Harnessing heat pump and steam recompression technology to meet the needs of industry 62
R. Gluckman, March Consulting Group, Windsor, United Kingdom

Impact of new technologies on future heat exchanger design 73
D. A. Reay, David Reay & Associates, Whitley Bay, United Kingdom

Energy recovery by mechanical recompression of hydrocarbon vapour . 83
J. P. Livernet, Société Rhone-Poulenc Chimie, Usine de Chalampé, France

Heat exchangers in plastic 88
J. Huyghe, General Manager, GRETh, Grenoble, France

Vapour compression in a brewery 98
E. Nolting, MAN Technologie GmbH, Munich, Federal Republic of Germany

Valorization of residual steam in brine evaporation 107
P. F. Bunge, Akzo Zout Chemie Nederland BV, Research & Technology, Hengelo, The Netherlands

Overview of the European Community research and development actions on low temperature heat recovery 116
P. A. Pilavachi, Directorate-General Science, Research and Development, Commission of the European Communities, Brussels, Belgium

SESSION IV: INDUSTRIAL PLANT—PROCESS CONTROL AND OPTIMIZATION

Control and optimization of processes 130
B. Kalitventzeff, University of Liege, Royal Military Academy, Belgium

An unconventional energy recycling project 141
H. P. van Heel, Managing Director of Hoechst Holland NV, Vlissingen, The Netherlands

The optimized process control of an ethylene plant 149
V. Kaiser and X. Hurstel, TECHNIP, Paris, France and S. Barendregt, PYROTEC, The Netherlands

Microprocessor system and digital regulation loops for increasing cowpers energy savings 159
A. Sciarretta, Process Control of Pig Iron Area at Italsider Taranto Steel Works, Taranto, Italy

SESSION V: ENERGY MANAGEMENT OF UTILITIES

New technics for the management of utilities in industrial plants . . 170
G. B. Zorzoli, Board of Directors, ENEL, Rome, Italy

Application of the SECI-MANAGER software to energy systems optimization and on-line industrial processes 178
M. Coeytaux, Serete Engineering, Paris, France

Energy savings and economic consequences resulting from the installation of a cogeneration unit (electricity—steam) at the Corinth refinery (Motor Oil Hellas) 185
A. Kalyvas, Motor Oil (Hellas) Corinth Refineries S.A., Athens, Greece

Software systems to optimize combined heat and power plant . . 192
J. Springell and D. Foster, Imperial Chemical Industries PLC, Billingham, United Kingdom

SESSION VI: ROUND TABLE ON FINANCING ENERGY EFFICIENCY INVESTMENTS

The financial engineering activity of the Commission 200
H. Carré and W. Faber, Directorate-General for Economic and Financial Affairs, Commission of the European Communities, Brussels, Belgium

Accelerating discrete energy efficiency investments through third party financing 204
D. A. Fee, Principal Administrator, Energy Saving Division, Commission of the European Communities, Brussels, Belgium

A new source of finance for investments in energy savings . . . 208
J. Junker, Bayerische Landesbank Girozentrale München, Federal Republic of Germany

Financing investment in energy efficiency from the manufacturer's point of view 211
P. Kalyvas, Motor Oil (Hellas), Athens, Greece

DISCUSSION AND CLOSING SPEECH

Discussion 214

Closing speech 219
M. Davis, Director, Directorate-General 'Energy', Commission of the European Communities, Brussels, Belgium

ZUSAMMENFASSUNGEN IN DEUTSCHER SPRACHE

Einführung 222

Prozeßintegration 223

Neue Techniken zur Rückgewinnung der Niedertemperaturwärme . . 226

Industrieanlagen—Prozeßüberwachung und Optimierung . . . 231

Energiemanagement in Energieversorgungssystemen 235

Podiumsdiskussion über Finanzierungsquellen für Investitionen zur Verbesserung der effizienten Energienutzung 237

RESUMES EN LANGUE FRANCAISE

Introduction 242

Intégration des procédés 243

Nouvelles techniques de récupération de la chaleur basse température . 246

Installations industrielles—Contrôle et optimisation des procédés . . 251

Gestion énergétique des utilités 255

Table ronde sur le financement des investissements d'efficacité énergétique 257

LIST OF PARTICIPANTS 261

INDEX OF AUTHORS 273

OPENING SESSION

Opening address by G. TURNER, Senator for Science and Research

Opening address by G. BRIGANTI, ENEA

Opening address by C.S. MANIATOPOULOS, Director General for Energy, Commission of the European Communities

OPENING ADDRESS

by
Professor George TURNER
Senator for Science and Research

Ladies and gentlemen,

On behalf of the Berlin Senate, I have great pleasure in opening this conference on "Energy efficiency in industry".

The people of Berlin are very grateful to the Commission of the European Communities for the considerable efforts it has made within and on behalf of Berlin, marking the 750th anniversary of the city by staging three international energy conferences here. The first conference in April was devoted to "Solar-heated swimming pools", in June there was "Coal in the heat market", and to conclude we now have this conference on "Energy efficiency in industry".

Increasing competition in both European and overseas markets has forced every entrepreneur to investigate all possible ways of cutting costs. The most obvious approach is to use energy efficiently, given that expenditure on energy is significant in virtually all fields. In this context, I feel that two of the issues to be discussed in this conference are of particular importance.

The first could be expressed as follows:

"How can process control be improved and techniques optimized ? What positive - or negative - experience has so far been gained ?"

The second major issue is that of funding:

"What sources of finance are available for improving the use of energy ? Who is entitled to request such investment aid for more efficient energy use, and how should this be done ?"

An answer to some of these questions may be found during this conference, and the Berlin Senate is hoping for some interesting tips based on experience gained throughout Europe.

Here in West Berlin, industry is a less significant energy user than in comparable Central European cities - representing only 13.5 % of total energy consumption. Nevertheless, any saving in energy is important since it cuts not only the energy costs of the company but also the pressure on the balance of trade; now an increasingly important consideration.

At the same time, the savings in primary energy help to reduce pollution in the city, particularly in the difficult field of air pollutants.

Here I would like to mention what we in Berlin see as another key issue, and one which will certainly become increasingly significant in the future, this being the generation of energy at the least possible cost to the environment. A decisive step has been taken here in Berlin. With the generous sponsorship of the Berlin electricity company "Bewag" - which contributed DM 4 million - a new department of "Energy conversion and environmental protection" has been set up at the Technical University of Berlin. This is the first step in developing a strong research basis in a field of such importance to conurbations such as Berlin.

In taking this action, it was the intention of the Berlin Senate to set an example by stressing the interrelationship between efficient energy conversion and the greatest possible protection of the environment.

Although there is a quite understandable desire to provide energy at the cheapest possible rates, the economic analysis must also include environmental protection or the expenditure necessary to restore a damaged environment.

In the future, we must invest more in our environment. We in Berlin have already made a start.

Although at present, and in such specific economic sectors as energy generation, environmental protection involves enormous costs, viewed in the long term and across the entire economy there is no conflict between ecology and economy. On the contrary: the development of new, more efficient environmental techniques is not just a scientific challenge. It opens up an international market that is, in view of its vital importance to future generations, in every sense "future-oriented".

We alone, however, cannot deal with the problems that arise. What is needed is intensive contact and cooperation between the scientists of every nation, since our environment is indivisible.

Nowhere in the world is better suited for reaching this conclusion, by "simply following one's nose", than Berlin.

If - not to put too fine a point on it - it stinks here to high heaven, then it is a transnational problem. The waters of the Spree and Havel, and particularly the air, are no respecters of artificial borders.

In a nutshell: if Berlin's air is to regain its once celebrated purity, joint action is the only alternative to failure. Common efforts to achieve efficient use of energy, and thus less pollution from energy generation, are a prime example of vital and feasible cooperation between both German states, and with other countries whose systems are different from our own.

Given the range of nationalities participating, we can expect not only a lively debate but also an exchange of views benefiting the countries involved. As a contribution to this process, there is the Senate's reception this evening at 6.30 pm, to which you are all most cordially invited.

I hope that we will all have a successful conference with many stimulating ideas, interesting discussions and valuable results.

OPENING ADDRESS

G. BRIGANTI
ENEA, Rome

I think this Conference is very timely, interesting and appropriate, and its results should have an impact that goes beyond the circles of specialists. I will try to qualify this statement.

The reduction of the energy content of the gross national products has been one of the three components of the strategic response of Europe to the energy crises, together with substitution of sources and the development of indigenous resources.

If we consider how this reduction has been obtained, we find out immediately that industry is the main responsible for the success of this policy.

However, when we try to interpret these data in terms of efficiency of energy use the task is not simple. There is a number of factors that interplay in this result. The aggregated figures that we are considering are actually the ratios between the energy consumption of industry and the added value of industrial production. This ratio has decreased not only because energy is used more efficiently in industrial processes, but also because the added value of industrial production has often increased. The value of industrial products is larger because they incorporate more technology, more design, more fashion, more response to individual tastes and requirements.

Another reason for the reduction of energy intensity in industry is the shift in the mix of products inside the industrial production. The market of basic goods that have a high energy and materials content is in many cases saturated, and their demand only covers replacement; demand for new goods goes toward more sophisticated, more "immaterial", more innovative products. In a general sense, development can be considered qualitative rather than quantitative: it concerns health, education, quality of the environment, free time, arts and therefore all the products that are instrumental to these objectives.

These two aspects of reduction of energy intensity are part of a general process of "dematerialization" which is common to all advanced societies; it has not been the consequence of the energy crises nor of the policies that were born of these crises, but the long term trend which was there has been accelerated by the energy crisis, through cultural evolution as well as economic pressure.

A third reason for the decrease of energy intensity in industry is perhaps less positive. It concerns the decrease in the productions that have high energy intensity, such as steel, plastics, fertilizers, accompanied by an increase in the import of such products (or by a decrease of previous exports). Such displacements of production from Europe to other countries (often developing countries) move the energy dependence from primary energy sources to energy rich materials. They may have positive connotations, such as a greater geo-political diversity of supply, or the possibility of cost reductions connected with the availability in some countries of very cheap energy; but it can hardly be regarded as a saving of energy: if the energy budget were to consider the energy content of imported and exported products (as in a way would be more correct) such a displacement would not result in net saving of energy.

The rest of the reduction of energy consumption by industry are linked with the concept of energy efficiency.

Still this reduction derives from a complex and composite panorama of different factors.

They include the elimination or reduction of energy wastage; the recovery and utilization of heat, the recycling of materials; the improvement of process efficiency through more accurate monitoring and appropriate control; the improvement of processes or the adoption of entirely new processes to obtain the same product; the subsitution of materials and other products to perform the same function or service.

Statistics tell us very little about the contribution of each of these factors to the reduction of energy intensity. Indeed, the trend towards goods with less energy content may be reflected by differences in the economic output of various industrial sectors, or by changes in energy consumption of these sectors. However, the analysis is shadowed by the effect of product shifts whithin each sector. For instance, very little sectoral shift appears in Italy, the United Kingdom and Ireland, where the changes of production occurred mostly within each industrial sector.

A better understanding of the mechanisms and of the opportunities of energy saving in industry is important for several reasons. One is to be able to predict in a better way the energy needs in the future. Another is to establish priorities in energy saving policies, incentives, investments, etc. Still another is to present industrial managers and investors with clear signals of what can be achieved and of the economic advantages associated with such policies.

It is from conferences like this that one can collect the basic material on which to base such assessment. The detailed analysis of interventions in specific industries, through process integration, heat recovery, process control optimization and energy management is a precious guide to ascertain results and opportunities of increasing energy efficiency in the industrial sector. The consideration of case studies makes the picture more concrete. Success stories make good example and provide guidelines for replication. Unfortunately, it is much less common to hear about failure stories, although we would have to learn just as much from them, in terms of mistakes to avoid as well as of obstacles to overcome on which to concentrate research and development efforts.

Much of the obvious to eliminate energy wastes and to use energy more efficiently through improved "housekeeping" practices has already been accomplished; the method of energy diagnoses by experts from outside and the preparation of energy managers inside industries have had a major role in bringing about these improvements. The EEC Commission estimates that there is still a great potential of energy reduction (of the order of 25 %) in continuing along these lines. Technical, legislative and organizational instruments to carry further this policy have already been identified and tested. It is now necessary to diffuse and implement this kind of interventions as much as possible. It may be useful to set up ad-hoc services that ensure capillary diffusion and replication of successful cases.

For the future, however, it is important to aim at deeper modifications, that involve prediction processes and renovation of plants. Energy saving in industry is therefore increasingly linked with innovation: it represents an opportunity to introduce new technologies, while at the same time every innovation in industrial processes opens the road to energy saving actions. The necessary investments are higher with respect to the energy management type of actions, but the results that can be expected go beyond the simple result of saving energy costs. They have to be judged in terms of quality of products, overall economic convenience, flexibility, protection of the environment, response to changing demand, etc. All these aspects are difficult to separate when considering process or product changes, and a comprehensive evaluation becomes mandatory.

It may seem trivial to tell this audience how important it remains to deploy the maximum effort in saving energy and using energy more efficiently in industry, no matter what the fluctuations of oil prices may be. I am sure that all of you share the feeling of the importance and of the strategic significance of the work we are all engaged in. Efficiency in energy use is economical, is profitable but its value goes beyond convenience: it is an essential part of a new model of society, a more sparing and more environmentally oriented industry; therefore it is also a model for less developed countries, whose way to development cannot repeat the wastage of resources and environment, the intensity of materials and energy of the traditional industrialization.

In this view, I wish this Conference the best success, also from Prof. Colombo.

OPENING ADDRESS

C.S. MANIATOPOULOS
Director-General for Energy
Commission of the European Communities, Brussels

The Commission of the European Communities is taking an active part in the celebrations marking the 750th anniversary of Berlin in order to demonstrate its firm commitment to the city.

Among the various Commission initiatives, I would mention the organization in 1987 of nine international conferences. As Director-General for Energy, I am pleased to say that three of these conferences were dedicated to energy.

Another three of the conferences were related to industry, concerning water resources, environmental protection and telecommunications.

The Commission has consistently stressed that Berlin is part of the European Community, and eligible to benefit from the opportunities available to every other region of the Community.

In this way it has given practical form to the declaration adopted by the six founding members of the Community on signature of the Treaty of Rome in 1957. In this declaration, which was ratified by subsequent new members of the Community, the signatories confirmed their solidarity with Berlin and their determination to contribute to its development.

Thus, for the purposes of regional policy, Berlin has the same status as the Community regions with an unfavourable location or which are geographically isolated.

In the field of technology, the Commission has been involved in a number of research, demonstration and investment projects. Taking the activities of my own Directorate-General only, these include the liquefaction and gasification of solid fuels, the construction of a more efficient district heating network, a pilot project concerning a large heat pump and the building of the Reuter West thermal power station.

In addition, two studies were launched in 1983 in collaboration with the Berlin Senate. The first concerns the creation of a data bank on the city's energy flow. The objective of the second is to devise a mathematical model for analysing energy problems in conjunction with environmental protection. It will be possible to apply the methods developed in these two studies to other Community regions. Finally, the Energy Institute of Berlin Technical University and the firm Innotec are cooperating with the Commission in organizing training and energy planning in China, the Asean countries and Morocco. Ten experts from Berlin are working in Community programmes for developing countries.

The conference on energy efficiency in industry fits in neatly with the Community's energy objectives for 1995 adopted by the Council of Ministers of the European Community in September 1986, and with the creation of a single market by 1992 decided by the European Council in 1987. Our energy objectives foresee a reduction of at least 20 % in energy intensity compared with 1985.

It is vital that this objective be attained, if our total energy consumption, which increases in step with general economic growth, is not to result in a growing dependence on external supplies.

Energy efficiency in the European Community increased by over 20 % between 1973 and 1983, while the share of imported oil in gross energy consumption was reduced from about 62 % to 44 % between 1973 and 1986. These achievements are largely the result of greater energy efficiency.

However, we at the Commission are not so vain as to believe that the Community can take all the credit for this success story. The period from 1973 to 1986 is characterized by major structural changes in European industry. Traditional industries such as steel and shipbuilding which are heavy energy consumers have declined, while new activities with a lower net energy requirement have made considerable advances. Substantial increases in oil prices between 1973 and 1986 also provided a powerful incentive for greater energy efficiency.

However, the results would not have been so spectacular without the efforts of the Member States and the supporting Community measures. At their recent informal meeting in Copenhagen and the Council meeting on energy, the Energy Ministers emphasized the importance of continuing to pursue our energy efficiency objectives. We are aware that the target of 20 % for 1995 is ambitious and will be difficult to achieve, because the process of industrial restructuring has not been completed and future price movements are uncertain.

Industry, while it has already made an enormous contribution to energy saving, will remain a priority sector because much remains to be done.

In 1985, Community industry accounted for at least 36 % of total consumption when energy as a raw material for the chemicals industry is included.

Compared with the buildings sector including heating and lighting of industrial premises which is responsible for 38 % of Community consumption, and transport with a 26 % share of consumption, industry is of prime importance in energy management at Community level.

As the second largest energy consumer, it is in industry's own interest to maintain and intensify its efforts to improve energy efficiency in order to make its products more attractive. It also has a duty to do so to contribute to the Community's independence in the energy sector.

Conversely, it is vital to the Community that industry does not falter in its efforts to make its prices more competitive. In addition, industry must forge ahead in the application of advanced technologies if it is not to lag behind in the international race.

It can be assumed that there is a considerable reserve of technological know-how that is still incompletely utilized and which could provide industry with the means of improving energy efficiency. The chief problem in applying these means is investment, particularly during periods of low oil prices.

The Commission believes that third-party financing is a possible solution to this problem. It recently organized an international symposium on this subject in Luxembourg, the results of which will be presented at the round table at the close of this conference.

However, the objectives of Community energy policy can only be achieved if Europe compensates for its shortage of primary energy by continuously promoting technological innovation. To this end, the Community is conducting a major research and development programme which has received funding in excess of 1.5 billion ECUs between 1984 and 1987, much of which has been dedicated to research on liquefaction and gasification of solid fuels, renewable energy sources, rational use of energy and environmental protection.

As a follow-up to the research and development work, the Community has carried out demonstration projects, some of which figure in the programme for the conference.

The idea of demonstration projects was born of the realization that success at the research and development stage did not always guarantee the success of a process or product on the market.

The transition from a research and development phase which has shown that an idea is technically and economically feasible to implementation on an industrial scale frequently involves technical and financial risks that act as a disincentive to entrepreneurs.

With the aid of the demonstration programme organized by the Directorate-General for Energy, which is organizing this conference, the Commission can bear part of the financial risks and smooth the way to the marketing stage. Since 1979, the Community has provided financial support of this type worth about 600 million ECUs to over 1 300 projects, 450 of them in the industrial sector. More than 300 projects concerning, among other things, renewable energy sources have already been completed, half of which, including 40 industrial projects, have been a resounding success. Of the 600 million ECUs concerned, almost 300 million ECUs were chanelled to industry.

Every successful project represents in itself a significant energy saving. However, our sights are set well beyond the actual projects. Our objective is to demonstrate the technical feasibility and economic viability of new procedures in the hope that their widespread adoption will lead to substantial energy savings at Community level. The replication of several steel projects, for example, has produced an overall saving of the order of 500 000 toe/year.

We attach great importance to the continuation of this programme after 1989, when the current 4-year period ends. We are devoting considerable attention to dissemination of the results in order to make the projects reproducible and to avoid duplication of effort. The Sesame databank is the main dissemination tool. It is now accessible to the administrations responsible for energy in all the Member States, and will soon be open to the public through the centres serving the european information market.

Information and the exchange of experience are additional objectives of this conference, which will doubtless represent an important step in the necessary progress of European industry to greater international competitiveness through technological development.

Before concluding, let me stress the quality of the speakers here today and the importance of the subjects discussed, which touch on vital aspects of the rational use of energy in industry.

I now have the pleasure of handing over to Professor Schäfer, Director of the Institute for Energy Management and Power Station Technology of Munich, who will give the first paper on fundamental ways of conserving energy and will then chair today's session.
Thank you.

SESSION I : OVERVIEW

Ways and techniques in the rational use of energy

WAYS AND TECHNIQUES IN THE RATIONAL USE OF ENERGY

By Prof. H. Schaefer, Munich

1. Introduction

The demand for energy to be used as sparingly as possible in order to preserve our environment and resources involves to some extent a contradictory states of affairs :

- Mankind must free itself from the environmental conditions by means of energy in order to achieve living conditions and a quality of life that can be regarded as humane.

- Any use of energy by man, even for his basic needs, has an effect on the environment and reduces the earth's resources through the consumption of materials, and use of space and fossil and nuclear fuels.

A first way of limiting factors affecting the environment is energy management, which, according to (1) is the sum of measures covering all activities designed to guarantee efficient use of available energy resources. These activities include energy saving, rational use of energy and substitution of energy sources for others, e.g. direct and indirect solar energy for fossil sources.

As shown in Figure 1, energy saving and rational use of energy are only synonymous at the point where they meet (market by "1"), i.e. in the area marked by less specific consumption of energy compared with a comparable state. This is achieved by reducing the specific consumption of useful, final and primary energy for the respective purposes and services. Taking the definition of rational use of energy in (2) as the use of energy by consumers in a way that is best suited to achieving economic aims - taking account of social, political and financial circumstances as well as environmental conditions - it includes the range marked by "2" in Figure 1, where extra specific consumption is brought about by additional energy services, e.g.

- a humane working environment;

- environmental protection technology;

- automation and mechanization; and

- overall optimization of work, materials, space and energy.

The higher specific consumption of energy arising out of these measures is justified by the increase in the quality of life, since in the final analysis energy demand is only one assessment criterion amongst many.

An example is flexitime. Everyone agrees that flexitime leads to more humane working conditions, but it involves a considerable extra outlay on energy since lighting and air conditioning are now needed each day some one and a half times actual working hours and it extends working hours into times of the day when the light of day and outside temperature are less than during the usual working hours previously.

Area 2 in Figure 1 contrasts with area 3 in terms of energy saving. In this area, less specific energy consumption is achieved by reduced demands for goods, services and comfort. This can be achieved by reducing the quality, quantity and range of goods and services, by lowering the room temperatures, by reducing lighting, by changing from individual to public transport and the like. It is difficult to draw a clear demarcation line between measures that contain a real sacrifice and measures that can be offset by non-energy steps. What is certain, however, is that not only the acceptance and social compatibility of the various supply techniques must be thoroughly examined but also energy saving measures themselves.

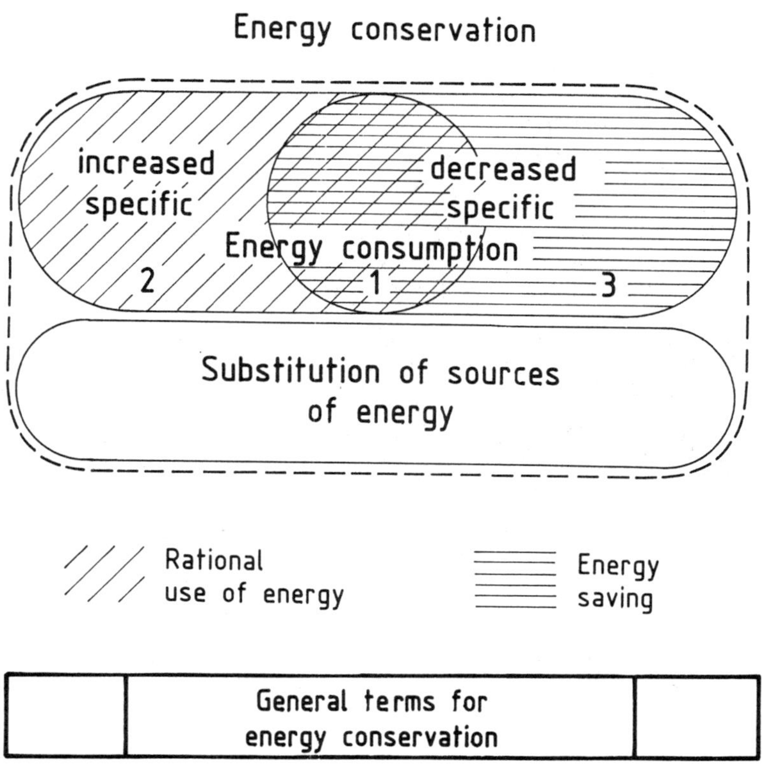

Figure 1

2. Energy analysis as a basis for rational use of energy

An absolutely essential requirement in any plans and measures to rationalize the use of energy is an analysis of the energy situation. An analysis of this kind, if it is to provide a suitable basis, must have the support of actual measurements, regardless of whether they are for individual installations, machines or entire plants. Fairly large areas (e.g. regions or entire countries) are statistically recorded in energy balance sheets.

Figure 2 takes the example of Germany to show how the final energy demand breaks down over the various consumer sectors and types of demand. The dominant role of heating, with a view to potential saving, is quite marked. Space heating at 35.4% and process heat at around 30% of the overall final requirements for Germany in 1986 take about two-thirds of the total energy used. Traffic accounts for 23.1% of the remaining final energy consumption while stationary power consumption in industry, the home and small consumers use around 10%. The proportionate share of final energy for lighting is approximately 1.8%.

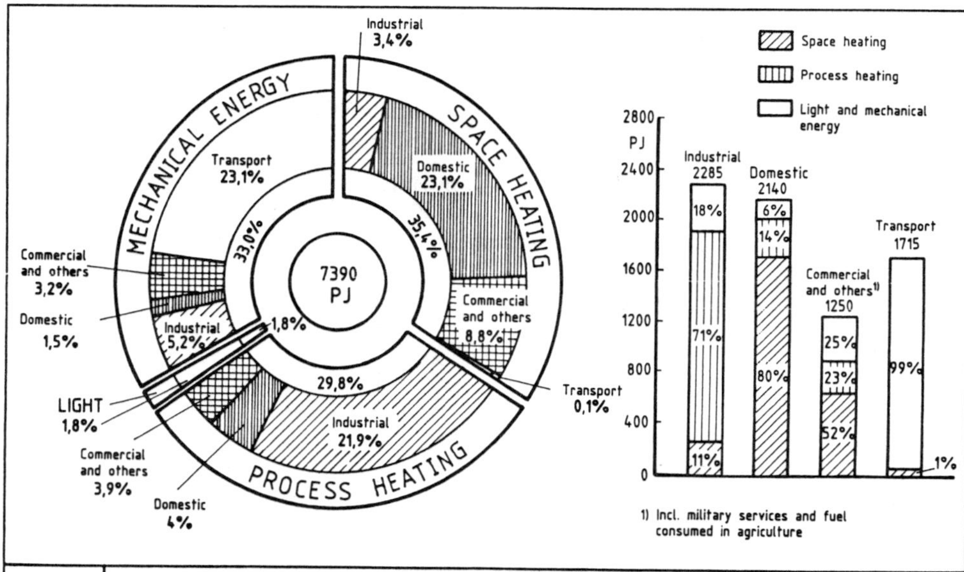

Final energy demand by consumer groups and types of requirement in the Federal Republic of Germany in 1985

Figure 2

The bar charts on the right hand side of the figure show the final energy consumption for industry, households, small consumers and traffic in absolute amounts and break them down percentage-wise into space heating, process heat and lighting and electricity.
This shows the dominance of space heating in households and amongst small consumers (almost 80 and 52%) and of process heat (71%) in industry.
Figure 3 gives an estimate of how the total use of final energy for process heat in industry is distributed over temperature ranges in steps of 100 K and over individual branches of industry.

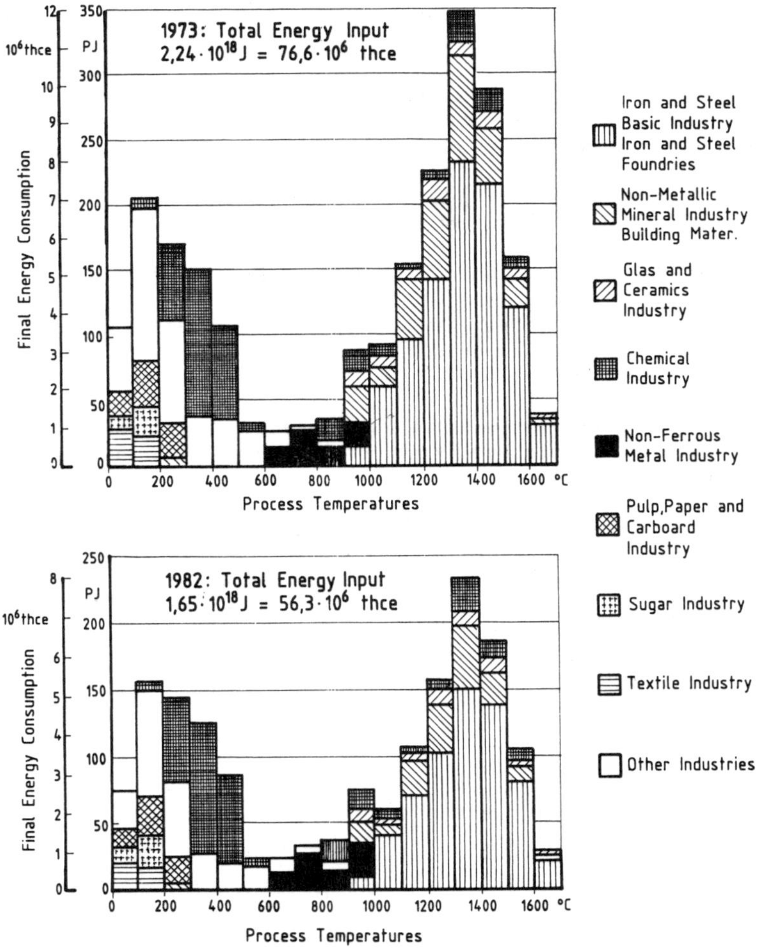

Final Energy Consumption for Industrial Process Heating in 1973 and 1982 by Groups of Process Temperatures

Figure 3

This is of course a relatively rough assessment based on knowledge of individual production methods and their specific energy requirements. The distribution curves for 1973 and 1982 show two peaks, the first in the temperature range around 200°C and the second between 1 300 and 1 400°C.

Overall results of this distribution had a marked effect on the iron and steel industry and the non-ferrous minerals industry. In 1973 both industries accounted for around 50% of total energy consumption. By 1982 this figure had dropped to well under 50%. Overall energy requirements for process heat in 1982 had dropped to 73% of the 1973 figure, this being caused by the reduction in the proportionate share of the primary industry.

An assessment of future trends cannot be made without an analysis of developments so far in energy consumption. In Figure 4 the specific fuel and power consumption is plotted against the net production index for the manufacturing industry in Germany. Whereas the specific fuel consumption dropped sharply over the period in question the specific power consumption rose slightly. A detailed analysis of the industry as a whole showed that trends in power and fuel consumption were down to three different effects.

The first effect is the activity effect, which gives an indication of what effect the changed production quantities, quantified as industrial net production index, will have on energy consumption.

The structure effect expresses to what extent changes in energy consumption can be explained by structural changes in the range of products.

The intensity effect illustrates how changes in power and fuel consumption can be the result of changes in specific energy consumption values. The results of an analysis of industrial final energy consumption in Germany, as carried out for the period between 1970 and 1983, are shown in Table 1.

	POWER		FUEL	
	s.o.	%	s.o.	%
Change in consumption	+32 679	+100	−135 326	−100
Activity effect	+17 835	+ 54,6	+104 720	+ 77,4
Structure effect	+ 6 805	+ 20,8	− 66 456	− 49,1
Intensity effect	+ 8 039	+ 24,6	−173 590	−128,3

Table 1 : Analysis of industrial final energy consumption 1970-83

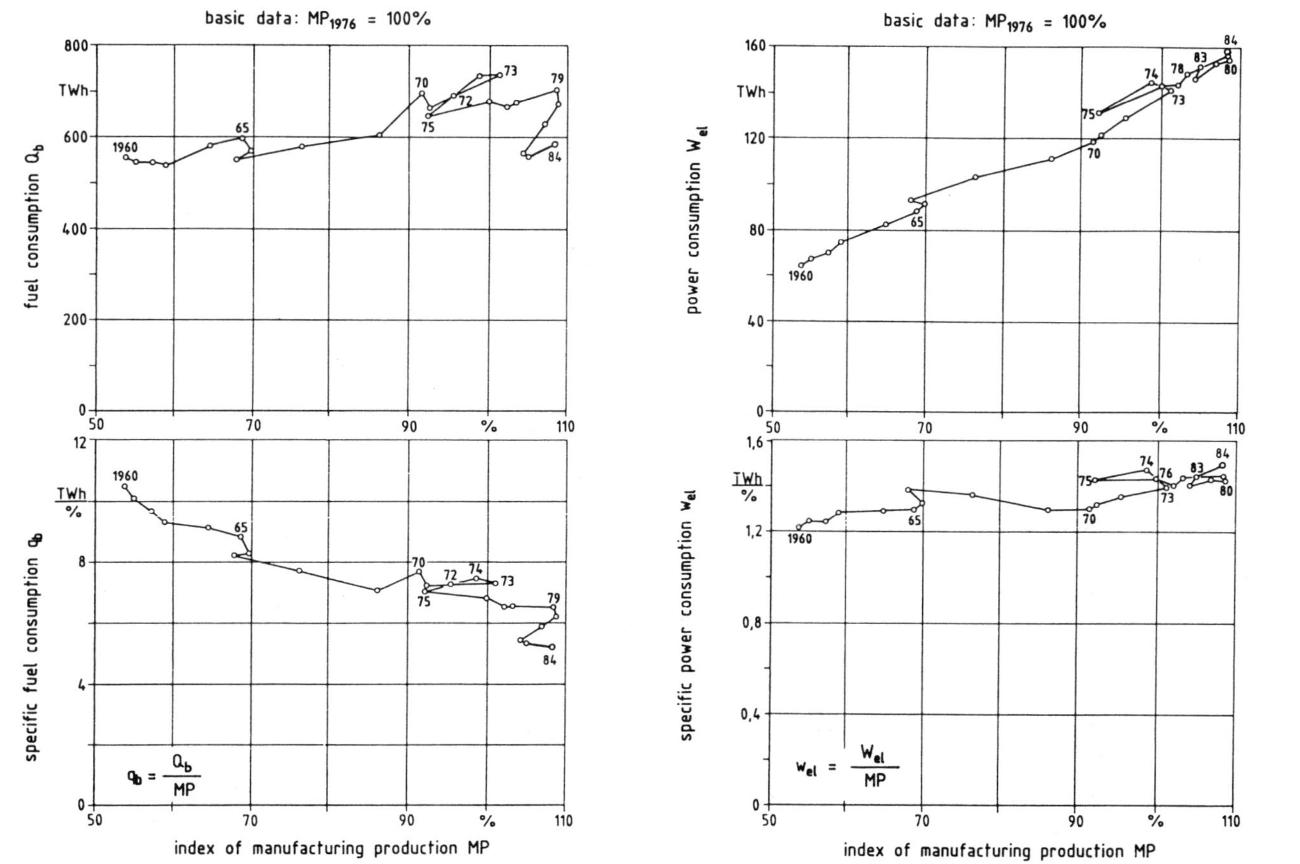

Fuel and power consumption versus MP of the manufacturing industry

Figure 4

Unlike electricial energy requirements, for which there is rising specific consumption, there is a sharp drop in the specific consumption of fuel. This is due not least to the rise in the use of electricity, which partly replaces fuel, but above all makes for more rational use of fuel by way of back-up energy in the system as a whole.

3. Underlying principle of rational use of energy

Basically there are five ways of using energy more rationally and more sparingly, these being :

- avoiding unnecessary consumption

- reducing the specific useful energy demand

- improving efficiency

- recovering energy

- using renewable resources of energy

3.1 Avoiding unnecessary consumption

All consumption is unnecessary which does not add to production or service or increases comfort. This includes, for example, machines and plant idling, overheating of rooms, water or other heat processes, excessively high pressure or quantities, etc. To avoid unnecessary consumption, technical measures such as dimmer switches for lights, limit switches on machines, compared with costs, can help to some extent; and instructions for the individuals using and operating energy-consuming machines can help a great deal.

3.2 Reducing the specific useful energy demand

Measures to reduce the specific useful energy demand are mainly of a technical nature and include heat insulation on all heating plant and optimum design and construction of the materials used in all manufacturing processes. For example, good aerodynamic design and reduced weight in aircraft lower the specific energy requirements for certain transport services. Choosing the optimum production process can also reduce useful energy requirements, e.g. gluing rather than welding, mechanical rather than thermal drying, non-cutting rather than cutting chaping, etc. This can be illustrated by the crankshaft of a middle-range passenger car. The volume to be cut in the wrought version compared with the cast version is 2.7 times and power consumption 1.8 times as high.

3.3 Improving efficiency

The efficiency achievable under normal operating conditions is often well below the nominal value of the machine because, on the one hand, energy consumption depends not only on the design of the machine but also, for example, on maintenance and, on the other, most machines consume a basic level of energy regardless of production, meaning that the specific consumption is a function of the load. High efficiency on production machinery can therefore be achieved by :

- energy-orientated design
- careful maintenance
- good load factor, i.e. proper dimensioning and optimum use of machinery in energy terms. The drop in specific energy consumption with increasing production is found more and more on modern machinery because of the incrase in mechanization, automation and regulation independent of load.

Machinery that consumes energy independent of load therefore has a high demand for energy when idling, and thus reducing the idling times in such machinery is of great significance to rational use of energy. Where possible, controls should be automated in such a way that unnecessary idling is avoided. As shown by studies by the "Research Institute for Energy Economics", a good 30 % of the total energy consumed during a shift by machine tools is usually down to idling during breaks and non-productive times.

A decisive factor for efficiency in power consumption is correct adjustment of the drive mechanism. Using electronic power-factor voltage setters the nominal efficiency and power factor can be achieved for practically any load between idling and rated load.

The prices of voltage setters, however, for drives under 10 kW are in the region of actuating drive costs. Proper dimensioning is cheaper. This applies in particular to pumps or ventilators when different throughputs are run. As can be seen in Figure 5, the power demand under part load drops only slightly by using the simple method of throttle regulation. Far better is speed regulation by electronic voltage and frequency setters. The costs of these devices, however, limit the profitability appreciably. This type of regulation only makes sense on pumps with changing loads.

It is easier and less expensive to avoid constant part load operating by correct dimensioning, although this requires the respective knowledge on the part of the machinery manufacturers and operators.

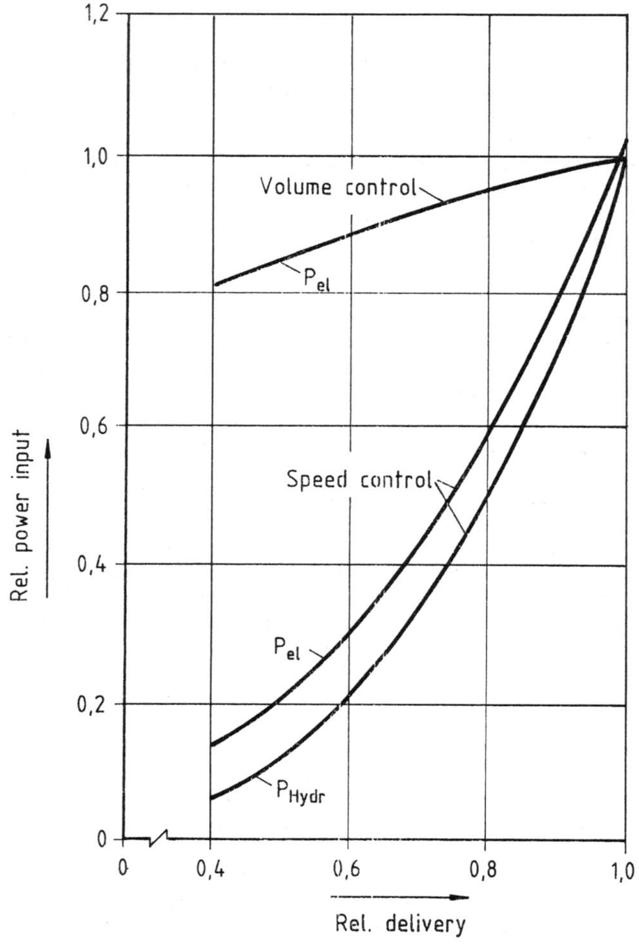

Relative power input of an electric driven pump

Figure 5

3.4 Energy recovery

Energy recovery in industry almost always means heat recovery. Economic use of waste heat is only possible if the waste heat emitted is concentrated, i.e. bound to one or few discrete substance flows (water, air, gases, solids), and not diffuse, generally in the form of large-area surface losses to the environment. The higher the temperature of the waste heat, the higher the energy content and the easier and cheaper the heat can be recovered. Figure 6 shows the temperature and type of industrial waste heat accrual in Germany in 1978. Almost half the industrial waste heat emitted is concentrated and thus one of the basic requirements for recovery is met.

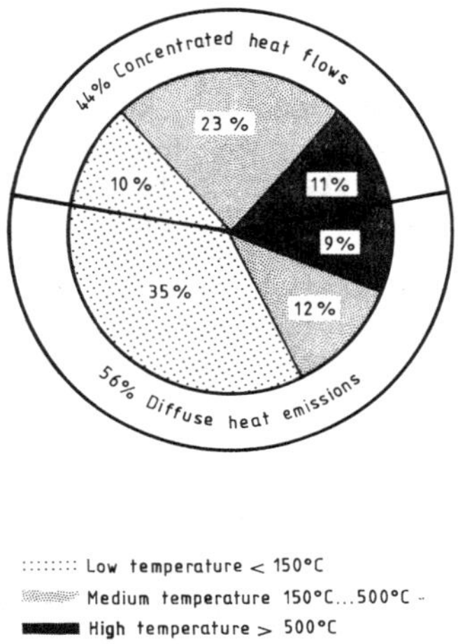

Waste heat in industry in the FRG,1978

Figure 6

Means of recovering heat are regenerative and recuperative heat exchangers and, at low temperatures, heat pumps. Nearly all systems of recovering heat require electricity as back-up energy, e.g. to transport the heat emission and take-up media, to control and regulate and, if needed, to drive heat pumps. In this instance, the extra consumption of electricity is a way of saving on heat consumption requirements. A prerequisite for using any kind of waste heat is a detailed analysis of the time and temperature profile of the waste heat in the processes producing it and of its potential consumers.

However, before trying to optimize a system through using waste heat, it would be better to reduce the waste heat by design and operating techniques to a technical and economic minimum. Nonetheless, the following priorities apply to the use of waste heat :

1. wherever possible, the waste heat arising in a given production process should be reused in the same process;
2. waste heat from industrial plant should, if possible, be reused in the same factory; and
3. only when the first two measures are exhausted should external use of industrial waste heat be considered.

3.5 Renewable sources of energy

In the long term, renewable sources of energy and the possible use of controlled nuclear fusion will be unavoidable for an economic supply of energy. This comes down both to the desire to preserve resources and to the ecological problems arising out of carbon dioxide emissions and, to a similar degree, steam emission.

Use of these sources of energy must take into account that compared with fossil sources of energy they are, because of their far lower power density, based on technologies requiring a greater supply of back-up energy, usually electricity. In addition, the space, surface area and material used per unit of power is generally greater than for conventional systems. The prospects of using renewable sources of energy for industrial production purposes must be regarded as slim in the more developed industrialized countries. In many developing countries, however, the chances are better because of more favourable climatic conditions and fundamentally different sets of circumstances.

4. Problems and limits of rational use of energy

Rational use of energy is not free. Even the great potential for using energy more efficiently through more sensible use of available means requires a considerable outlay for the requisite information process and with it the change in the individual's approach.

Since the ratio between expenditure and use often differs greatly from measure to measure, each case should be carefully examined to find the optimum way of rationalizing the use of energy, with particular attention paid to the fact that several measures taken in one and the same area can have a strong influence on one another. In general, priority should be given to measures with the highest savings efficiency even though they may not be very spectacular and in practice call for greater vigilance on the part of those involved. However, through these cheaper measures, energy saving can often be achieved to such a degree that capital intensive projects, e.g. heat recovery plants, are not needed.

Furthermore, the technique used is generally more complex, less clear and the action principles increasingly impenetrable. Maintenance and repairs call increasingly for specially trained and qualified personnel and thus become a far greater cost factor than previously.

It is especially important that every aspect affected by a measure to rationalize the use of energy is taken into account. These aspects may be of an economic, ecological, social and even human nature.

References

(1) Energy Terminology
A Multi - Lingual Glossary, 2nd Edition
CEC London, UK, Pergamon Press

(2) Grundbegriffe der Energiewirtschaft und Energietechnik
Erarbeitet vom Ausschuss "Terminologie in der Energietechnik"
der VDI - Gesellschaft Energietechnik
Berichterstatter : Prof. Dr. - Ing. H. Schaefer
Sonderdruck aus Brennstoff-Wärme-Kraft 32 (1980) Nr. 8,S.334/37

(3) Schaefer, H., Wege und Techniken zur rationelleren
Energiebedarfsdeckung. FfE Schriftenreihe Nr. 1, Feb. 1981

SESSION II : PROCESS INTEGRATION

Energy savings in the manufacture of crankshafts - an example of integrated analysis based on detailed measurements

Process integration using pinch technology

Process integration in a benzole refinery

The results of a process integration study to improve energy efficiency at a British brewery

ENERGY SAVINGS IN THE MANUFACTURE OF CRANKSHAFTS - AN EXAMPLE OF INTEGRATED ANALYSIS BASED ON DETAILED MEASUREMENTS

Dr. M. RUDOLPH
Professor of power production and power-station technology, Munich

In order to quantify potential energy savings in specific cases, relevant data must be measured and then combined to yield an analysis of the energy-consumption patterns, under normal operational conditions, of the plant concerned.

This paper examines two production lines for the manufacture of crankshafts in a car factory. Both lines comprise a large number of single-purpose machine tools, each carrying out one of the sequence of machining steps required to turn the original unworked piece into a finished part - which then undergoes hardening.

Figure 1 provides an overview of the rated power of all drive units. There is a major difference in the degree of linkage and hence in installed power. In the "production sequence", only six of the 34 machine tools are linked together and only one station is loaded automatically. All other transport and loading is carried out by hand, using hoists or so-called "dog-bar" conveyors. On the "production chain", by contrast, there is full hydraulic linkage of all 23 machine tools, with the workpieces being both loaded and transported automatically.

Although the total cumulative power rating of all main spindle drives is almost the same for the two production lines, there are considerable differences in the figures for individual types of machine tool.

The installed power of lathes and boring machines is greater in the production chain than in the production sequence. The opposite is true of milling machines and grinders.

Far less installed power is devoted to the movement of slide units in the production chain than in the production sequence. To some extent, this is balanced by power-rating differences in the machines' hydraulic system. The pumps used to wash the machined crankshafts are 50 % more powerful in the production chain. There is no significant difference in the total power of the other drive units (e.g. chip transport, supply of coolant or lubricant).

Although it is possible to obtain the details of the installed power of drive units either from the factory inventory or from the power-rating plate on each motor, energy consumption must be measured.

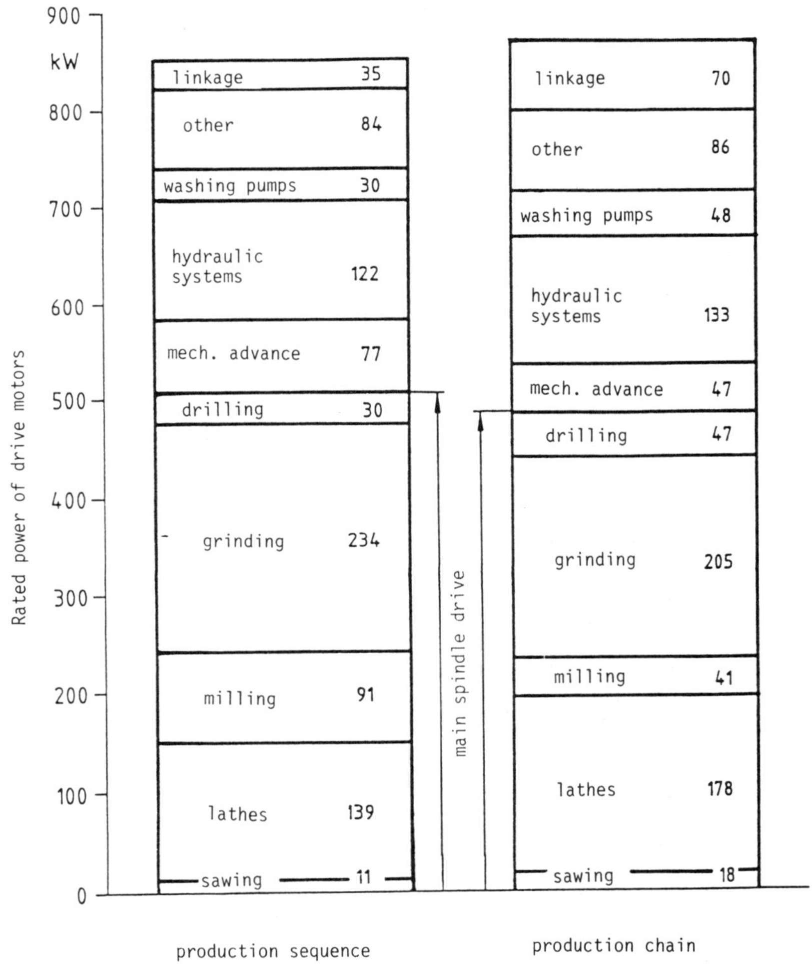

Fig. 1	Breakdown of installed power of drive units in two production lines	1607 87

In the case of long production lines, this may be extremely time-consuming, since it is not enough to simply measure total consumption at a central power-supply point - even if one exists. Any assessment of possible energy savings at individual stations requires separate measurements at each one, these comprising not only the consumption of electricity but also the use of a chart recorder to measure variations in instantaneous power (power during machining and idle power) over a number of load cycles. It is advisable to do the same with individual main spindle drives.

A comparative analysis should then be made of power consumption in the two production lines. Figure 2 illustrates power consumption per finished crankshaft, broken down by machine type. The consumption figures are then further subdivided into consumption during actual machining time and other consumption. It is only for the linkage units that this kind of breakdown is impossible.

If one then looks at the energy consumption of the linkage units, the 1020 Wh recorded for the production chain is more than three times the figure for the production sequence - hardly surprising given the greater degree of linkage. The comparison of the rated power of these linkage installations, revealing a ratio of only 2 : 1, indicates that far too powerful motors were chosen for the partial retrofitting of the production sequence linkages. Indeed, the recorded power charts showed that consumption at load at some of these hydraulic units barely exceeds idling levels.

If consumption by the linkage units is subtracted, the total energy required to finish a crankshaft is approximately 8 500 Wh or 5 800 Wh in the production sequence and production chain respectively. In the context of the virtual equality in total power rating, the obvious conclusion to be drawn is that much too powerful units have generally been installed in the production sequence. This assumption is apparently strengthened by the fact that the capacity of the production sequence is approximately half that of the production chain ($26\ h^{-1}$ as against $50\ h^{-1}$).

The situation looks rather different, however, when one considers that the two lines are machining crankshafts differing not only in weight but also in the material and technique used to finish the unworked piece (production sequence : CK 45 Forging; production chain: GGG 60 casting). The key difference, however, is in the volume of material to be removed during machining which, at 946 cm^3, is 2.7 times higher for the production sequence than for the production chain. If energy consumption at stock-removal stations (saws, furning, milling, grinding and boring machines) is related to the volume of material to be removed, the 11.2 Wh/cm^3 of the production chain is over 40 % higher than the production sequences figure of 7.9 Wh/cm^3.

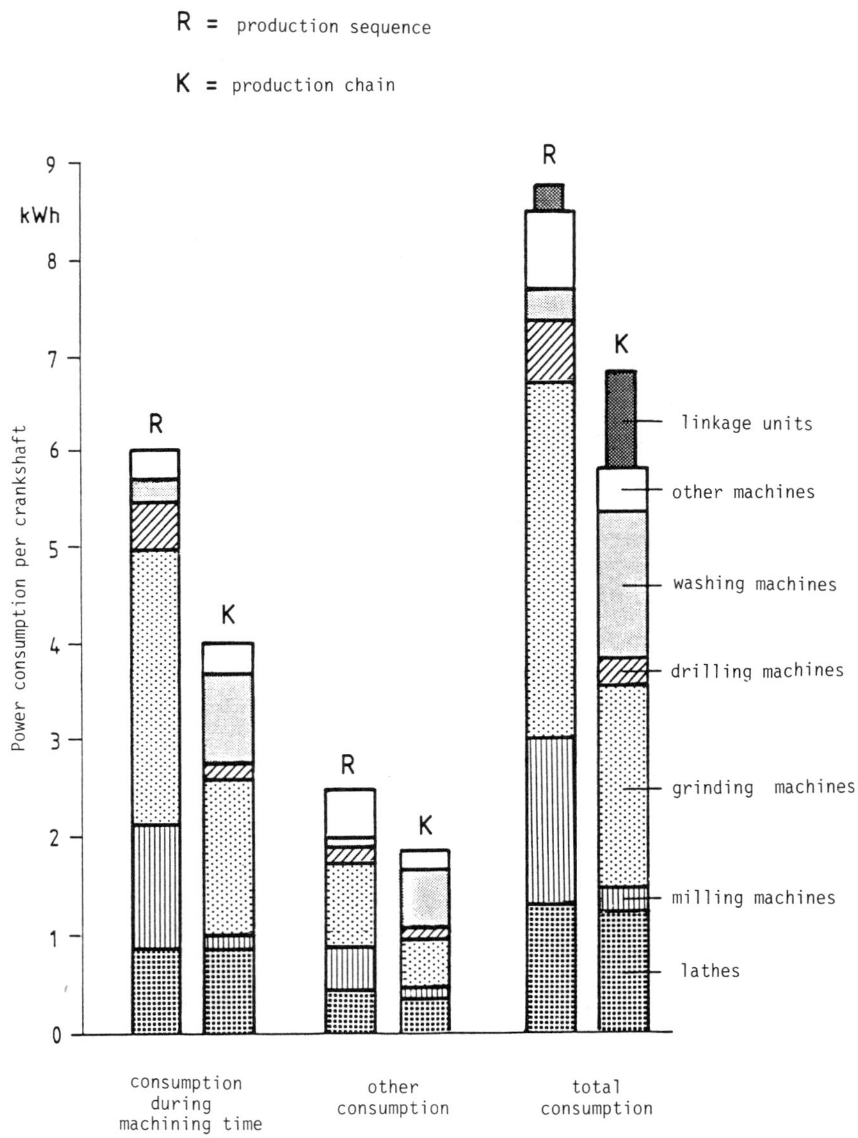

Fig. 2	Breakdown of power consumption in two production lines	1606 87

Temporarily disregarding the load factor and consumption by auxiliary drives, the explanation must be sought in the varying energy requirements of individual machining techniques. Accordingly, the energy consumption of the main spindle drive was measured on some of the stations of both production lines at a range of machining rates. These measurements can be found in Figure 3. The specific energy consumption as a function of the stock removed tends to fall at higher stock-removal rates. At the same rate, the specific energy consumption of a lathe will be roughly twice as high for a casting as for a forging. Other things being equal, stock can be removed with a lathe using only a fraction of the energy needed for grinding or milling.

Measurement of the stock removed at each station would provide more detailed information about individual energy requirements. This was, however, not possible during this research. In consequence, the only aspect of machining technology that can be put forward to explain the above difference in specific energy consumption is the generally higher level of energy required to machine spheroidal graphite cast iron (GGG) rather than hardened steel (CK). This is obviously a more important consideration than the fact, apparent from Figure 2, that it is precisely the more energy-intensive machining techniques of grinding and milling which account for a greater proportion of energy consumption in the production sequence than they do in the production chain.

The major cause of this disparity is that two manufacturing steps carried out on lathes in the production chain are carried out on grinding and milling machines respectively in the production sequence. Recourse to lathes in these instances would allow energy consumption during actual machining on the production sequence to be cut by at least 1100 kW per crankshaft. No further attention will be paid to any other potential substitution of energy-intensive steps, since this would involve the discussion of manufacturing and organizational details outside the scope of this research.

Figure 2 also contains a diagram breaking down consumption figures into actual machining and other time, and hence offers an initial means of identifying where energy could be saved by switching off drives when workpieces are not being machined. For cutting tools, the actual machining time is that during which material is being removed from the workpiece. In the case of other manufacturing units, the actual machining time must be defined in such a way that it comprises the work directly required to achieve the aim of that manufacturing step.

On this basis, the actual machining time and associated energy consumption were determined for each individual station. The ratio of this consumption to consumption during an entire machining cycle (i.e. including non-machining time) is defined here as the "output utilization".

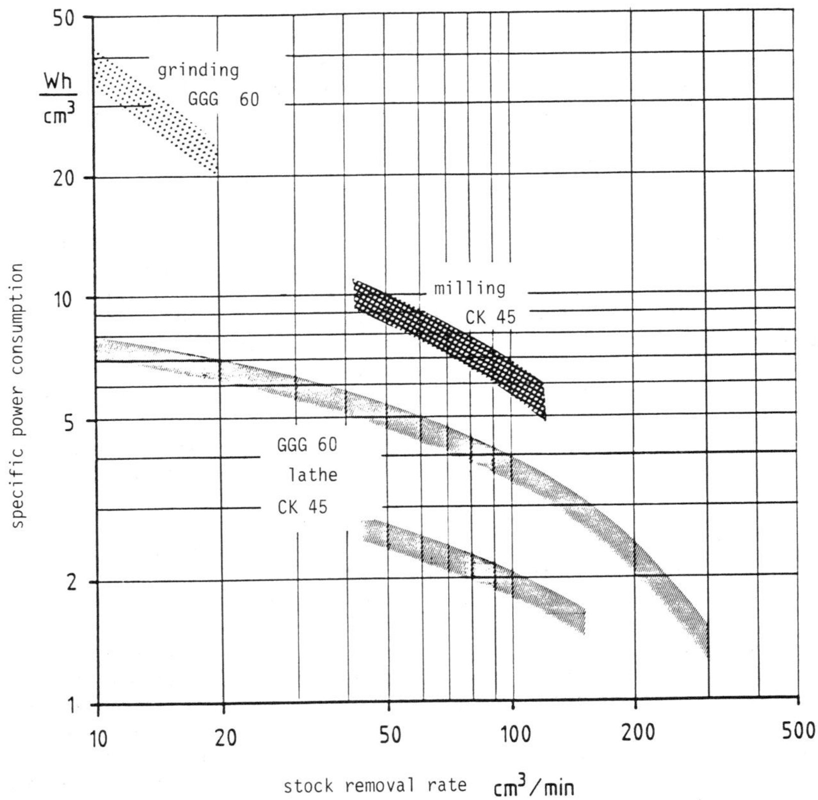

Fig. 3	Specific power consumption of main spindle drive of machine tools	1605 87

In the production sequence as a whole, output utilization is 71 %, only negligibly higher than the production chain's 68 %. There is strikingly low output utilization, only 60 %, at the wash station on the production chain. If the wash stations in both production lines are disregarded, the lines show the same overall output utilization of 71 %. In general, energy-intensive stations in both lines are characterized by above-average output-utilization figures.

It would not, however, be realistic to assume that all the non-machining consumption shown in Figure 2 can be cut. For example, it is normally necessary to maintain the oil pressure of hydraulic drives throughout the operating time. Most other auxiliary drives also have no idle consumption in the usual sense. The greatest potential for switching off machinery during non-machining time is offered by the main spindle drives of cutting tools and the wash-pump drives of the wash stations. In most cases there would be no difficulty in modifying the control programme accordingly. Since it was possible during this research to carry out a separate measurement of the main drive at only a few stations, insufficient data are available to predict potential savings.

If energy is to be used rationally, the rated power of the motor must be appropriate for the maximum drive power required. Particularly where the load varies over time, as we found at most machines, unfavourable partial-load performance arising from over-powerful motors was clearly reflected in excessive energy consumption. On the basis of the measurements made, the load situation can be described using the following parameters :

- the maximum load factor, expressed as the ratio of the maximum power consumption experienced to the power consumption during operation at normal rating, and

- the average load factor during machining time, expressed as the ratio of the average power consumption during machining time to the power consumption during operation at normal rating.

The maximum output utilization was found to be 64 % for both production lines. This total value was not calculated on the basis of the total output of each production line, but rather by totalling the individual maxima of the various stations. Most stations have a maximum load factor of between 30 and 90 %. When assessing these findings, it must be borne in mind that the maximum power consumption, particularly of the main spindle drives, is not always constant - not even for special-purpose machine tools for serial production. Indeed, a number of factors are involved, such as the diameter of the grinding wheels, blunting of cutting tools or tolerances in the dimensions of the unmachined part, as is apparent from Figure 4. Power consumption in hydraulic pumps is very strongly influenced by the oil temperature.

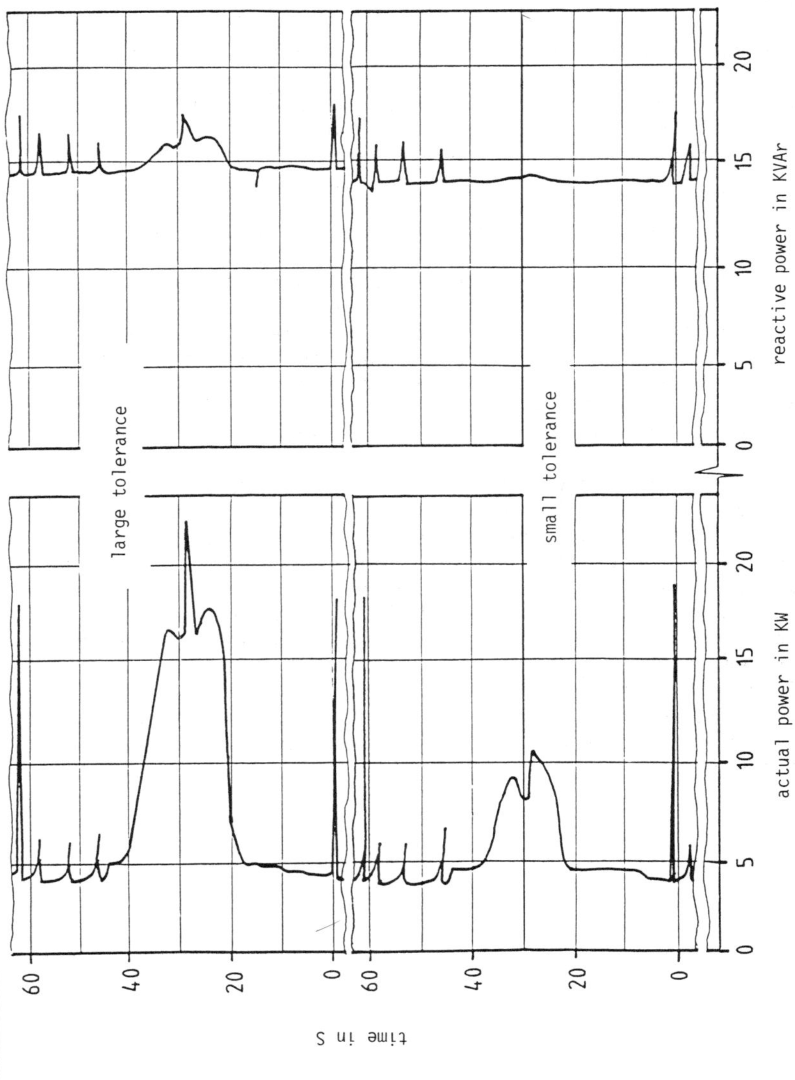

Effect of tolerance on the power consumption of a lathe

Fig. 4

1604
87

The average load factor during machining time was rather lower on the production sequence (30 %) than on the production chain (34 %). Leaving aside some exceptional cases, all stations registered values of between 20 and 65 %.

Expressing the two load factors as a ratio yields the utilization factor for machining time, i.e. the ratio of average power to maximum power. This provides some indication of the average load level to be expected using a motor of optimal power rating based on the recorded load factors during machining time.

Utilization factor values of 47 % and 53 % were calculated for the production sequence and production chain respectively. It is noticeable that particularly low values, between 30 and 40 %, were recorded at certain grinding and milling stations - precisely those at which consumption is higher. Here in particular, savings can be made by avoiding over-powerful drive motors.

The associated potential energy savings must, of course, not be overestimated. For example, a rough calculation based on optimizing the main spindle drive on a milling machine, where the maximum load factor is approximately 40 % and the average load factor during machining time only 20 %, indicates potential energy-consumption savings of approximately 13 % during actual machining time. Such a situation was, however, rare since the high-consumption drives in particular were relatively well suited to their role and yielded high load factors, exceeding 80 % in some cases.

The above discussion is devoted exclusively to electricity consumption during the mechanical machining of crankshaft blanks, together with some important factors influencing the level of consumption. The approach would, however, be incomplete (and maybe even misleading) were it to neglect differences in the prior manufacturing steps producing the unmachined pieces. As an approximate comparison, the production of a forged blank requires the consumption of a good 15 kWh more electrical energy than is the case for a casting. In other words, this difference is virtually 8 times the difference between the consumption figures of the two production lines and is a further point in favour of the production chain since, in the above case, it processes castings.

PROCESS INTEGRATION USING PINCH TECHNOLOGY

B. LINNHOFF
Centre for Process Integration
UMIST, Manchester, UK

and

A. EASTWOOD
Linnhoff March Ltd
Manchester, UK

1. INTRODUCTION

Pinch Technology has proved effective in developing optimal integrated process designs for both new plant and retrofits. This has been demonstrated in hundreds of successful projects, carried out mainly in the UK, the USA and some European countries. These projects have covered a wide range of industries using both continuous and batch operations.

Basic research and development of pinch technology has been carried out by the Process Integration group at UMIST, Manchester, U.K., headed by Professor Bodo Linnhoff. The work has been supported for a number of years by an international consortium of companies including Exxon, BP, Shell, BASF, Union Carbide, ARCO, Dow, M.W. Kellogg, and others.

As an example of how pinch technology is applied, consider Figure 1. The design shown is based on a recent case study. The evaporator plant on the left consists of a multi-effect system with a feed pump driven by a back-pressure steam turbine. Thus, the evaporator plant in itself represents a total energy system in which low pressure exhaust steam from the turbine is used for process heating in the evaporator. This appears to be an apparently well integrated CHP scheme with little scope for improvement - but is it?

It is important to review the performance of the evaporator plant not in isolation but in the context of the overall process. Some salient features of the remainder of the plant are also shown in Figure 1 and we shall see later how pinch technology helps us to easily improve on the design shown here.

2. REVIEW

2.1 Energy Targets, the 'Pinch', and Minimum Total Cost

The first stage in any application of pinch technology is to represent the entire process on a temperature-enthalpy diagram by composite curves, as shown in Figure 2. These curves represent the cumulative heat sources and heat sinks in the process. Their construction is fully described elsewhere (1).

For a specified minimum heat transfer driving force, ΔT_{min}, the composite curves define utility heating and cooling targets.

The point at which ΔT_{min} occurs between the composite curves is termed the pinch. The pinch divides the process into two thermodynamically separate systems, each of which is in enthalpy balance when the utility targets are applied (Figure 3). It follows that the energy targets will only be achieved if there is no heat transfer across the pinch. This is the pinch principle, based on fundamental thermodynamics (2). It can be summarised as follows:

A	=	T	+	XP
Actual Energy Consumption		Target Energy Consumption		Cross-pinch Heat Flow

The implication is that the target energy consumption can only be achieved if cross-pinch heat transfer is avoided. We have obtained a simple but fundamental design rule (1)!

Figure 4 shows that an increase in ΔTmin increases the energy targets (ie higher energy cost), but also provides larger driving forces (ie lower capital cost). This relationship can be quantified by calculating capital cost targets from the composite curves (3).

The annualised capital and energy cost targets can then be combined, as shown in Figure 5, to indicate the optimum value of ΔTmin. Thus, pinch technology addresses both thermodynamics and economics (4).

3. EVAPORATOR CASE STUDY

We can now go back to the evaporator case study introduced in Figure 1. When we analysed the total process at the site we constructed a composite curve similar to that shown in Figure 6.

It was immediately apparent that the turbine exhaust steam, which condensed above pinch temperature, was used for the evaporator heating duty below the pinch. This represented cross-pinch heat transfer.

In the context of the overall process, to be efficient the evaporator should be heated by process heat recovery below the pinch. The composite curves showed that suitable heat was available from one or both of the distillation column condensers.

Next, the turbine exhaust steam is free to perform suitable heating duties which should be above the pinch. Again, the composite curves showed that suitable heat sinks were one or both of the distillation column condensers. The overall revised configuration is shown in Figure 7. There is a genuine saving in hot utility (HP steam) to the process. The turbine now rejects heat above the pinch and is said to be 'appropriately placed' (5,6), see Figure 8.

This example demonstrates the need to consider total systems. The original designers had 'integrated' the turbine with the evaporator - but in isolation. It is possible to design a whole series of part-systems, each individually optimised, but end up with a total system which is, as a whole, non-optimal. Pinch technology allows us to always look, with little time required, at the overall system both for new and existing designs.

3. GENERALISED

3.1 Utility Integration

The grand composite curve (Figure 9) is an alternative representation of the process on a temperature-enthalpy diagram (1). It describes the profile of the net process heating and cooling requirements which need to be met by external utilities.

It is easy to determine the overall most suitable choice of external utilities from the grand composite curve (1). For example, in Figure 10 the utility heating target is met by a combination of flue gas and steam, and the utility cooling target is met by cooling water and refrigeration. The temperature and duty of each utility has been chosen to provide a good match between the utility profiles and the grand composite curve.

It is important to note that the grand composite curve does not represent a process as designed but an optimised process. If the process is designed optimal the curve represents it. If it is not, the curve represents what could be rather than what is. In a retrofit study this could be extremely important. Not only can heat recovery be optimised but appropriate process modifications can be identified to suit the available utilities. The identification of appropriate process modifications is an essential part of any pinch technology study (7). The following example, based on another plant study, is chosen to illustrate this point.

4. CHEMICAL PLANT CASE STUDY

Figure 11 illustrates the relevant section of a chemical plant. Feed is preheated by fractionator overheads before passing to the main reactor. Exothermic heat of reaction is removed by hot oil which is used to reboil the adjacent stripper. Hot utility, QH, is applied to the fractionator reboiler by means of 3500 KPa steam from the central boiler house.

There was an overall incentive on site for power generation and conventional examination by inspection had shown that waste heat from a heat engine could be used to replace the existing 3500 KPa steam to the fractionator reboiler. Since the process temperature was 220°C, however, the choice of heat engine was limited to a gas turbine. The steam to the reboiler could be replaced by hot turbine exhaust gas as shown in Figure 12.

The annual savings for this project was £340,000 for an installed capital investment of approximately £1 million, equivalent to a three-year payback. The savings represented a 37% reduction in total energy bill and the project was considered marginally viable. However, Pinch Technology was consulted prior to a final decision to check for missed opportunities.

We can see how the gas turbine shows up in terms of pinch technology by constructing the profile for the turbine exhaust gas over the process grand composite curve (Figure 13). The construction immediately reveals that the "utility" heating supplied by the exhaust gas is greater than target and is available at higher temperatures than necessary.

Further, the grand composite curve shows a large process sink at a relatively low temperature. If we can 'open up' the system so that this low temperature sink is available to the external heat engine we should be able to improve the engine efficiency dramatically. To do this, we need to remove the 'notch' in the grand composite curve by modifying the process in some way, see Figure 14.

Pinch Technology uses rules which enable beneficial modifications to be identified (7,8). For the present case, these rules lead to an increased pump-around flowrate for the hot oil from the reactor. This will effectively upgrade the heat in this stream (Figure 15) and remove the 'notch' in the grand composite curve (Figure 16). External hot utility (ie turbine exhaust gas) is now only required at a temperature of about 150°C. This is equivalent to steam at 5 bar and, clearly, we have re-introduced the option of a steam turbine. However, remember that any grand composite curve shows the process as could be, not necessarily as is. We still need to reconfigure the process heat recovery. The final retrofit (total project) is shown in Figure 17.

Annual energy savings were slightly less than with the gas turbine project at £300,000, but the capital cost was only £250,000 (ie down by 75%!) giving a simple payback of less than one year (9).

This simple example highlights several important aspects of pinch technology. We have seen that it is possible to tailor the utilities options to a given heat and material balance. We have seen that it is possible to "adjust" the process heat and material balance to suit. Furthermore, these considerations are carried out in the composite curves, grand composite curves, etc, not by means of cumbersome flowsheet or design evolution. It is always possible to finally turn the "chosen option" into a design.

Using conventional design methods the designer would identify potential for improvements by inspection in the flowsheet. Invariably, this design will "start" from the existing process and the existing equipment arrangement. This makes it very difficult to spot and "undo" the limiting features (eg pump-around flowrate).

Pinch Technology looks at the basic process, unfettered by the constraints and penalties inherent in the existing arrangement. Consequently it is able to provide insights into the absolute potential of the system as a whole. The designer now has a systematic approach which gives a clear understanding of the interactions within the process, between the process and utilities, and between the utilities options available. With the procedure summarised in Figure 18, we can systematically ensure optimisation of the overall system as an integrated whole.

5. TRACK RECORD

This technology has now been used in more than 500 industrial applications worldwide. Sucessful commercial applications have involved both continuous and batch processes. The cost benefits, in terms of both energy and capital, have been dramatic (see Table 1).

The savings in energy costs have been particularly marked. In continuous processes the application of Pinch Technology, rather than traditional design methods, on average result in energy savings of around 25% of total energy consumption. In retrofit projects savings of between 15 and 30% at a one year payback are a common result.

With retrofit projects involving batch processes the savings can be significantly higher. Industrial case studies have demonstrated savings of 40-70% of energy consumption and cycle time improvements are common.

In the UK applications have been made in a wide range of industries: petroleum, general chemicals, petrochemicals, pulp and paper, food and drink, cement, steel, pharmaceuticals, and fibres. A part of this success is due to the significant effort made by the Energy Efficiency Office of the UK Department of Energy. They have published a series of reports covering applications they have sponsored (10).

Industrial experience in the running of integrated plants has shown that initial fears that the integration can result in plants that are difficult to start up and to run are unfounded. In fact integrated plants (provided the integration is carried out correctly!) are often better than traditional designs in both respects.

Techniques for the engineering of flexible plants are now emerging (11). These show that, again contrary to expectation, integrated plants can be more flexible than unintegrated ones. If properly designed they can cope with changes in operating conditions very well. Through integration,flexibility can be achieved with smaller overdesign margins than hitherto thought possible,for good integration places the margin where it is used most effectively (11).

Again these findings are being supported by industrial experience. In one recent case study,a plant that had to operate under twelve different conditions was studied. The result was a project that made substantial energy savings (at the specified two-year payback). Not only were the 12 operating scenarios satisfied but a significant debottlenecking of the plant was also achieved.

6. CONCLUSIONS

Pinch technology has a proven record of industrial application. This covers a wide range of industries using both continuous and batch operations. Savings in both energy and capital have been substantial.

The technology is based on the pinch principle which is founded on fundamental thermodynamics. Since targets can be set for both energy and capital cost, the best capital-energy trade-off is known and can be achieved.

Since the technology presents a clear picture of the whole system, it enables the engineer to focus on the absolute potential overall. Integration of apparently complex systems becomes a manageable task. Beneficial process changes are spotted easily.

Recent work has highlighted another important aspect of well integrated plants. Contrary to expectation they can be more flexible than their less integrated equivalents.

TABLE I

Process	Type of Project	Energy Savings £/annum	Capital Cost £	Payback months
Petrochemical	Retrofit	700 000	330 000	<6
Speciality chemicals	Retrofit	93 000	38 000	5
Speciality chemicals	Retrofit	55 000	4 000	<1
Inorganic bulk chemical	New	160 000	savings	-
Speciality chemical	New	50 000	saving: 75 000	-
Organic bulk chemical	New	400 000	same	-
Bulk acid	New	40 000	saving: 70 000	-
Organic bulk chemical	Retrofit	670 000	400 000	7
Edible oil	Retrofit	450 000	-	-
Whisky distillery	Retrofit	300 000	-	10-15
Synthetic resins (Batch-multipurpose)	Retrofit	250 000	-	3-24
Ethylene	New	430 000	saving:470 000	-
Oil refinery	Retrofit	1 875 000	3 000 000	19

7. REFERENCES

(1) LINNHOFF, B., et al., "User Guide on Process Integration for the Efficient Use of Energy", Institution of Chemical Engineers, London 1982.

(2) LINNHOFF, B., "Pinch Technology for the Synthesis of Optimal Heat & Power Systems". ASME AES Vol. 2.1, pp 23-35.

(3) TJOE, T.N., and LINNHOFF, B., "Using Pinch Technology for Process Retrofit", Chem. Eng., April 28, 1986.

(4) LINNHOFF, B., and AHMAD, S., "Supertargeting. Optimal Synthesis of Energy Management Systems", paper presented at the ASME Winter Meeting, Anaheim, December 1986, ASME AES Vol. 2.1, pp 1-14.

(5) TOWNSEND, D.W., and LINNHOFF, B., "Heat and Power Networks in Process Design", Part 1: 'Criteria for placement of heat engines and heat pumps in process networks', AIChE Journal, 29, No 5, pp 742-748 (1983).

(6) TOWNSEND, D.W., and LINNHOFF, B., "Heat and Power Networks in Process Design", Part 2: 'Design Procedure for Equipment Selection and Process Matching', AIChE Journal, 29, No 5, pp 748-771 (1983).

(7) LINNHOFF, B., and VREDEVELD, D.R., "Pinch Technology has Come of Age", Chem. Eng. Prog., pp 33-40, July 1984.

(8) LINNHOFF, B., and PARKER, S.J., "Individual Process Improvements in the Context of Site-Wide Interactions", I.Chem.E. Annual Research Symposium, Bath, 1984.

(9) EASTWOOD, A.R., and LINNHOFF, B., "CHP and Process Integration". Presented at the 51st Autumn Meeting of the Institute of Gas Engineers, London, November 1985.

(10) Reports available from:
The Energy Efficiency Office
Thames House South
Millbank
London SW1P 4QJ

(11) KOTJABASAKIS, E., and LINNHOFF, B., "Sensitivity Tables for the Design of Flexible Processes (1) - How much Contingency in Heat Exchanger Networks is Cost-effective?", Trans. I.Chem.E., Vol. 64, pp 197-211, 1986.

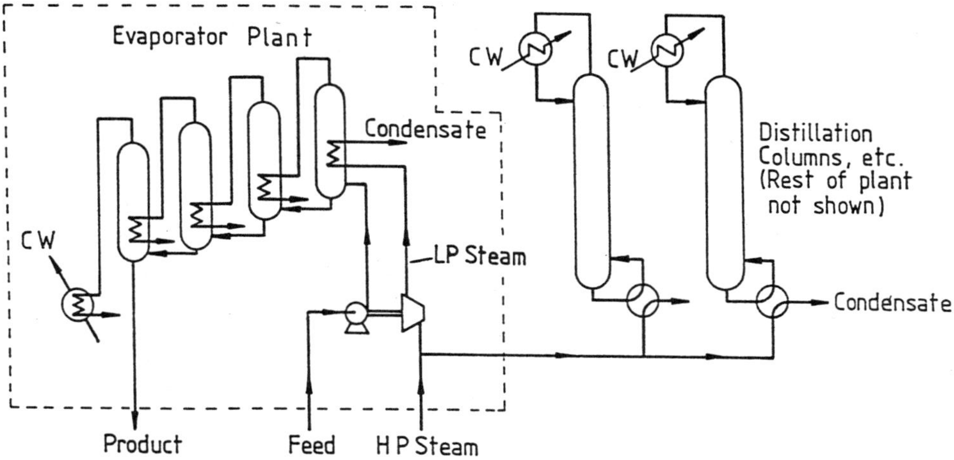

Figure 1

An efficient CHP system ?

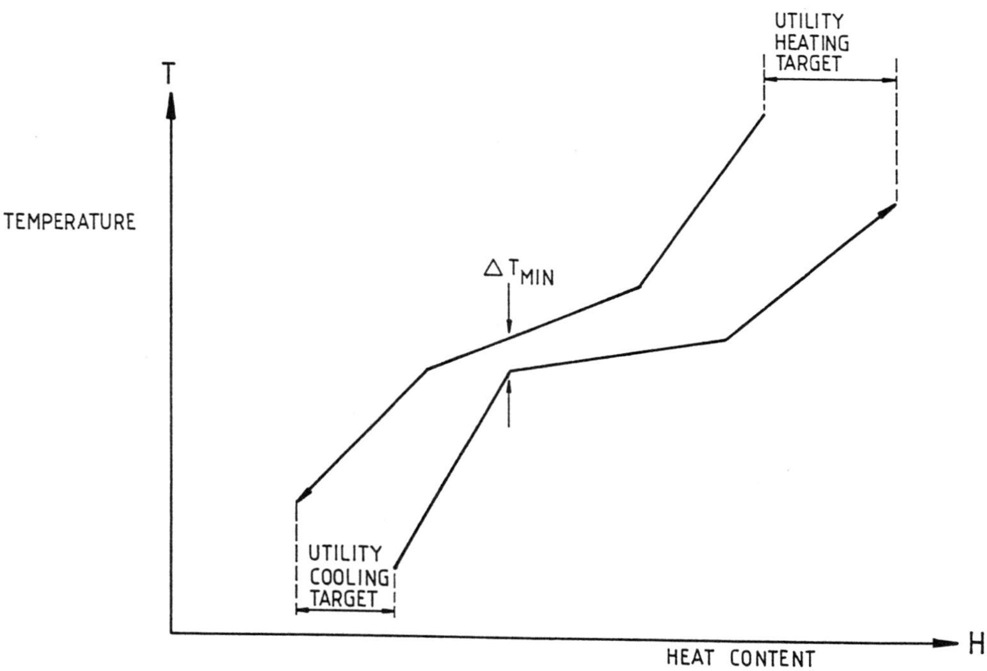

Figure 2

The composite curves

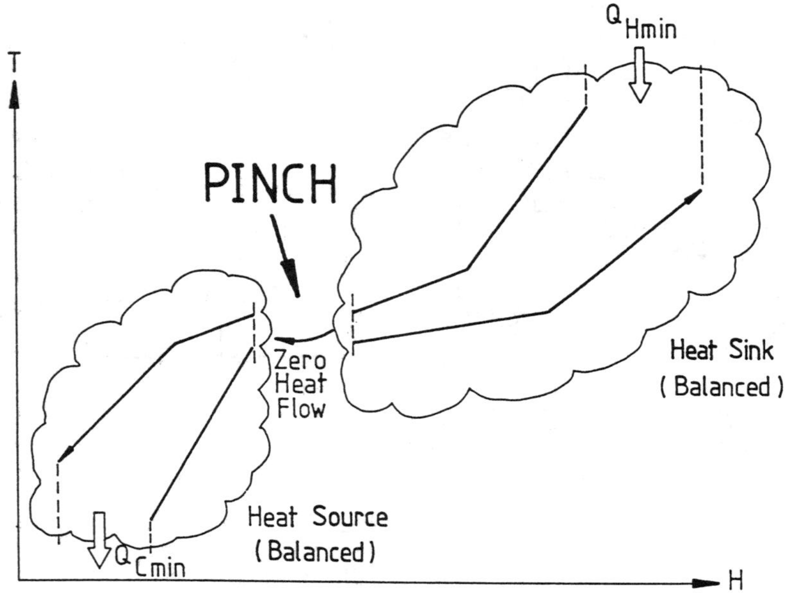

Figure 3

Two systems are defined by the Pinch

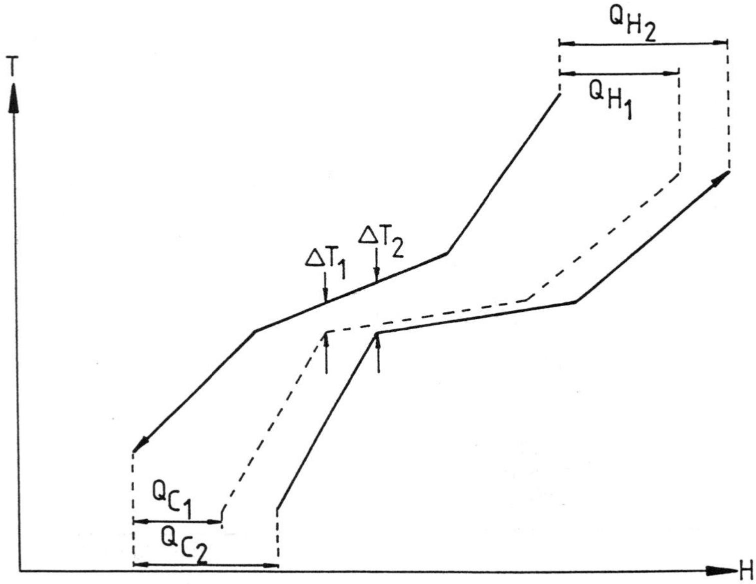

Figure 4

Energy, capital trade-off

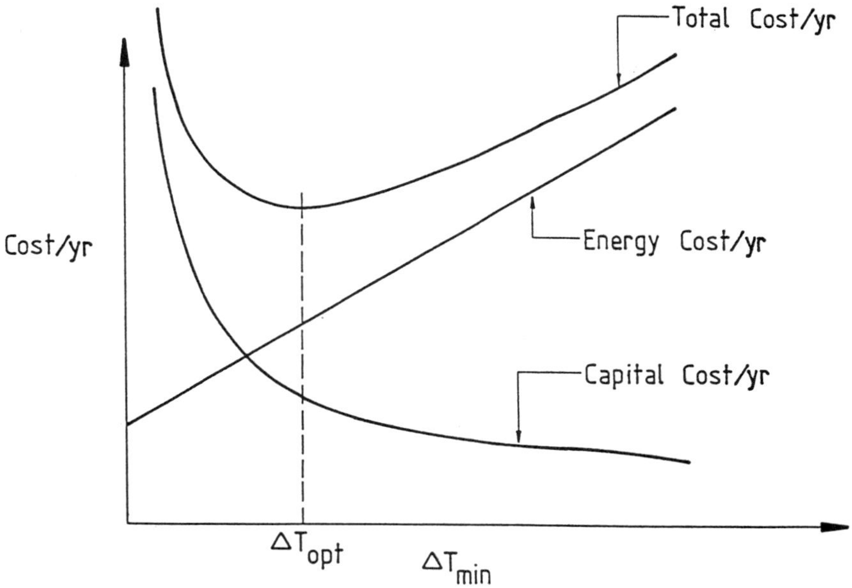

Figure 5

Optimisation before design

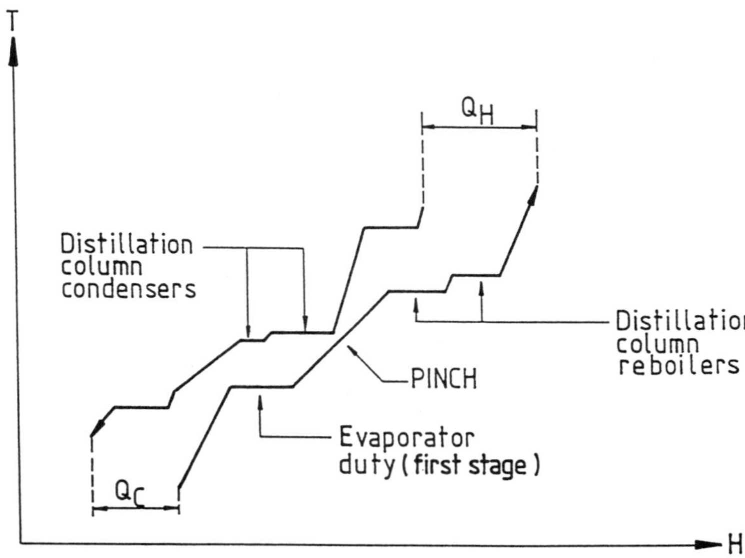

Figure 6

Composite curves of overall evaporator plant

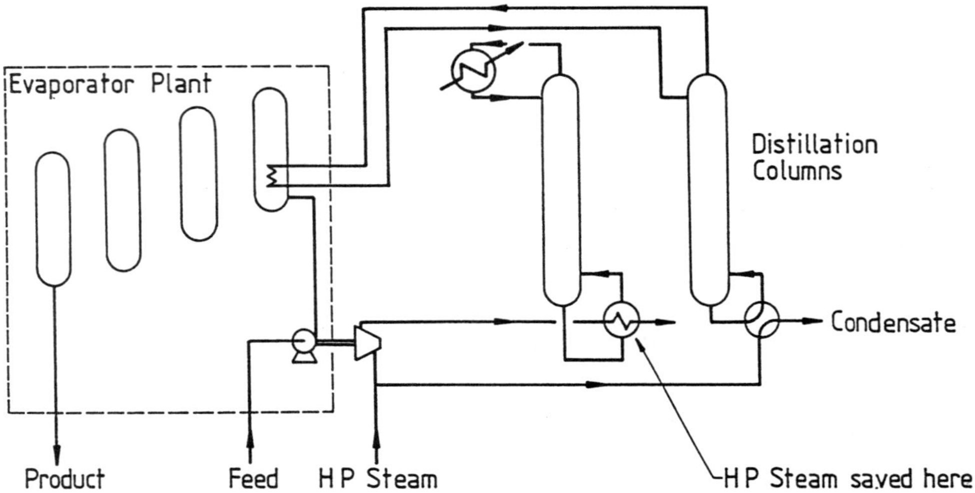

Figure 7

A better CHP system !

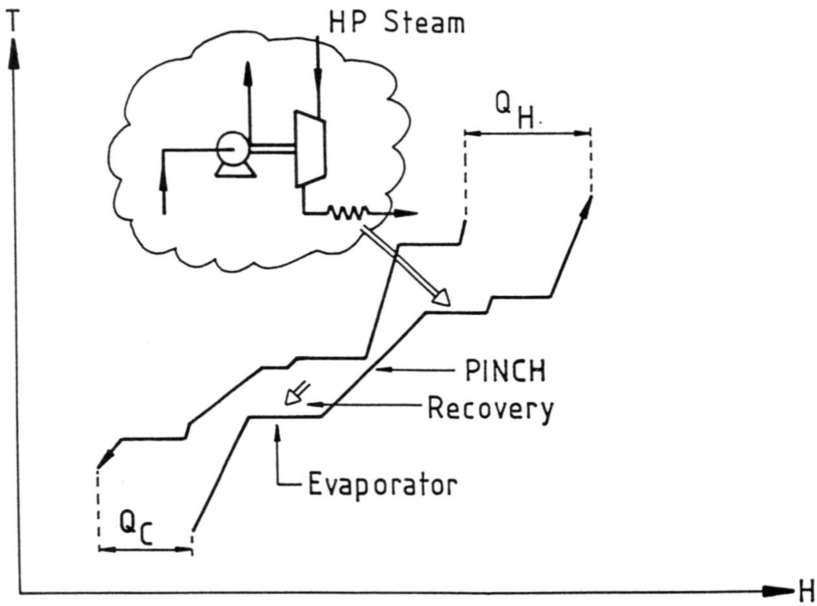

Figure 8

The better system in terms of Pinch Technology

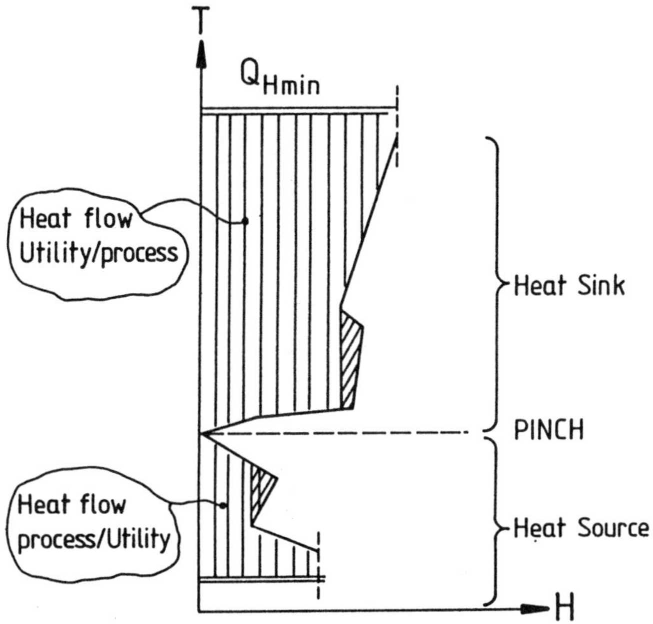

The GRAND COMPOSITE CURVE

Figure 9

Concentrating on the Process/Utility Interface

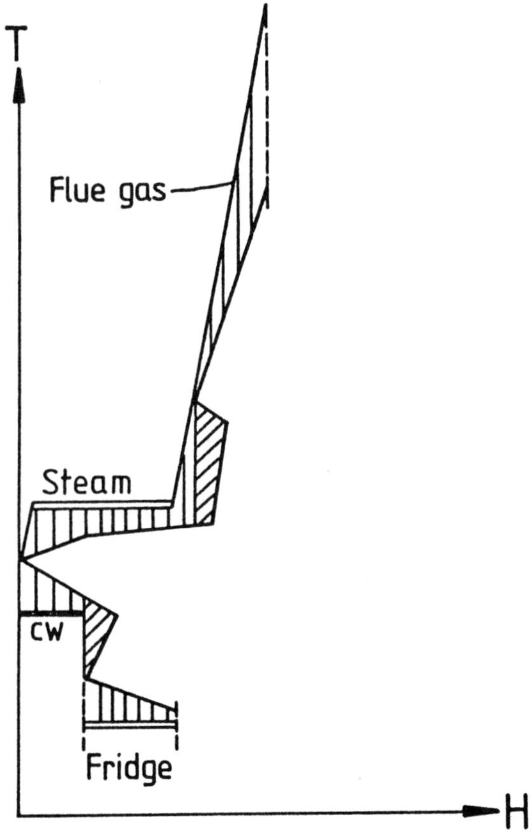

Figure 10

Mixed utility targeting

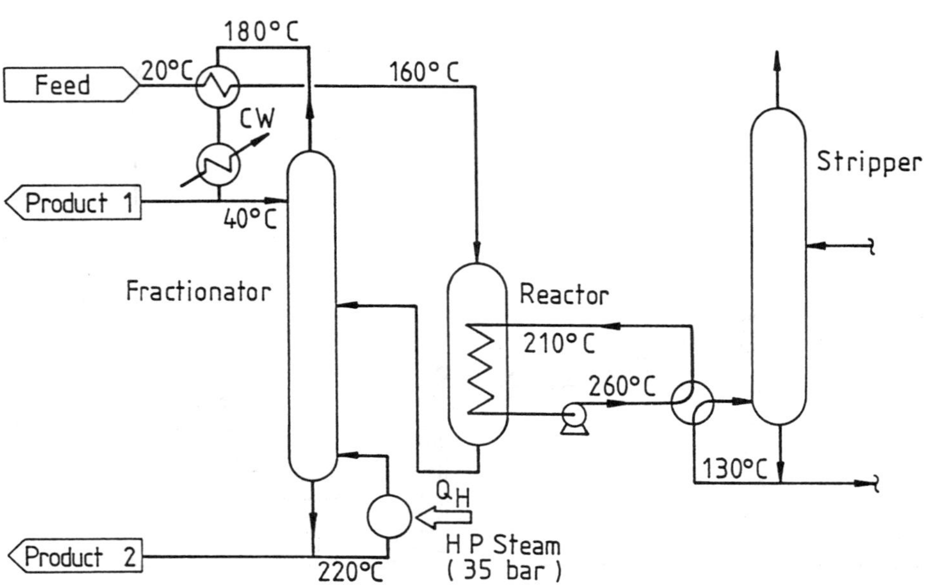

Figure 11

Part - Flowsheet for Chemical Plant

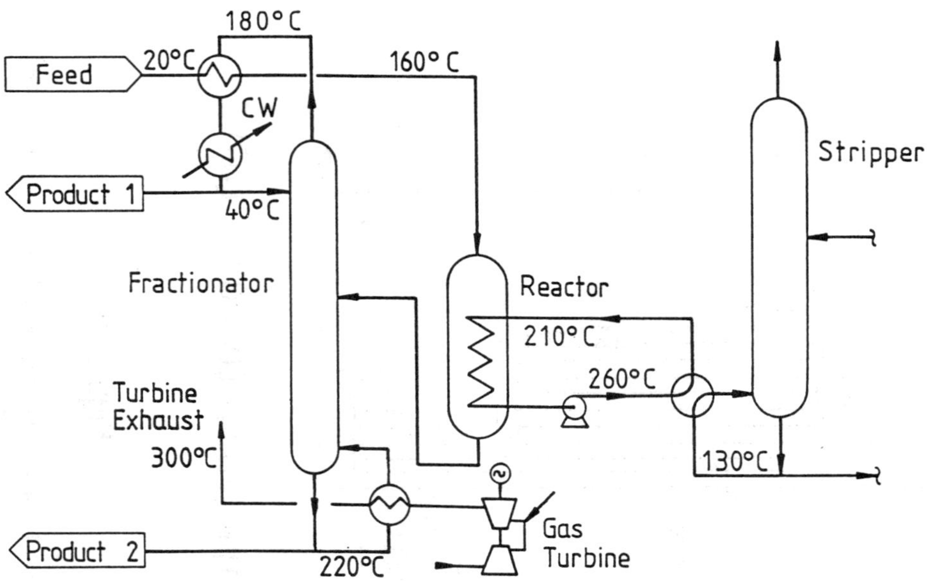

Figure 12

Initially proposed CHP scheme

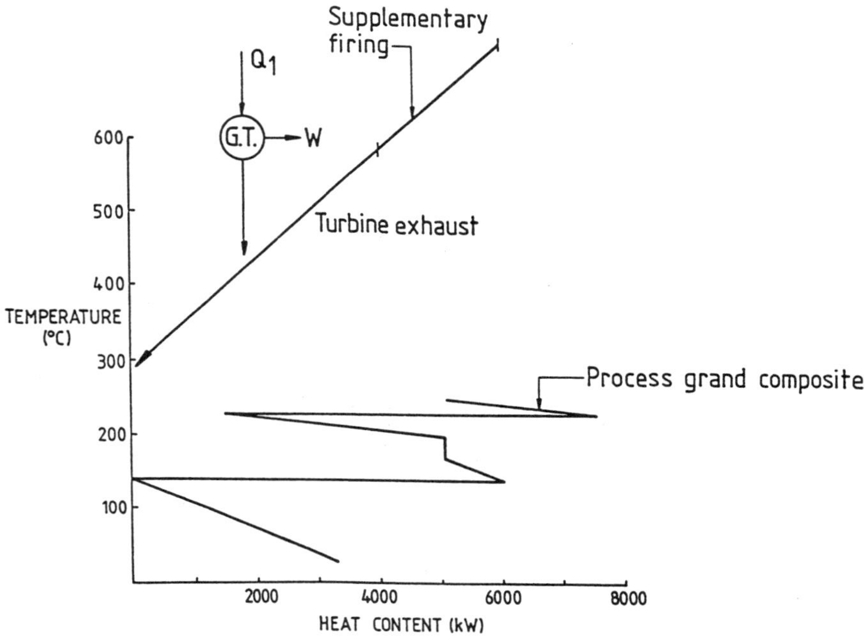

Figure 13

Pinch analysis of initial scheme

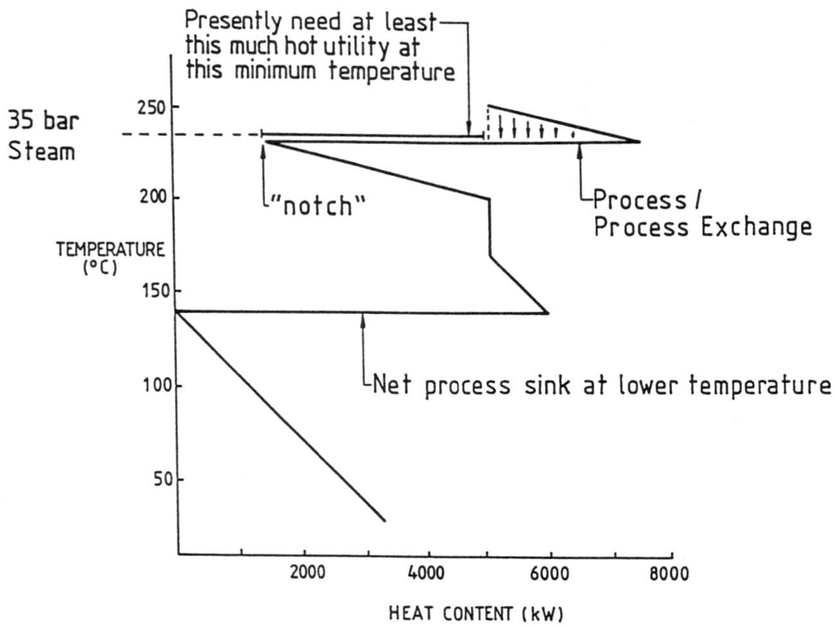

Figure 14

Identifying the limiting process feature

What if ?

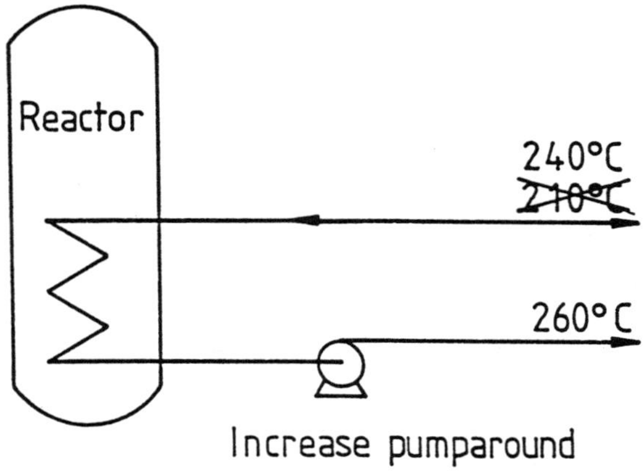

Figure 15

The heating process change

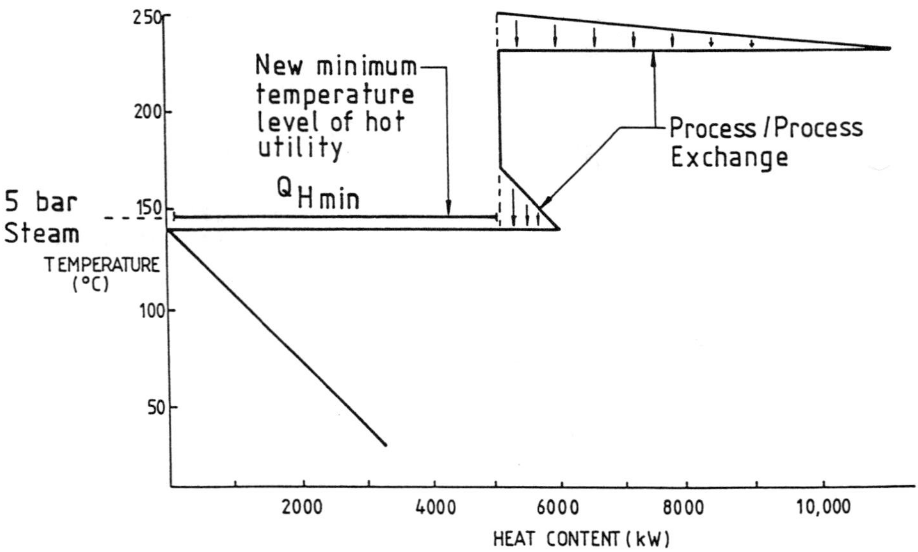

Figure 16

Ready for a better scheme

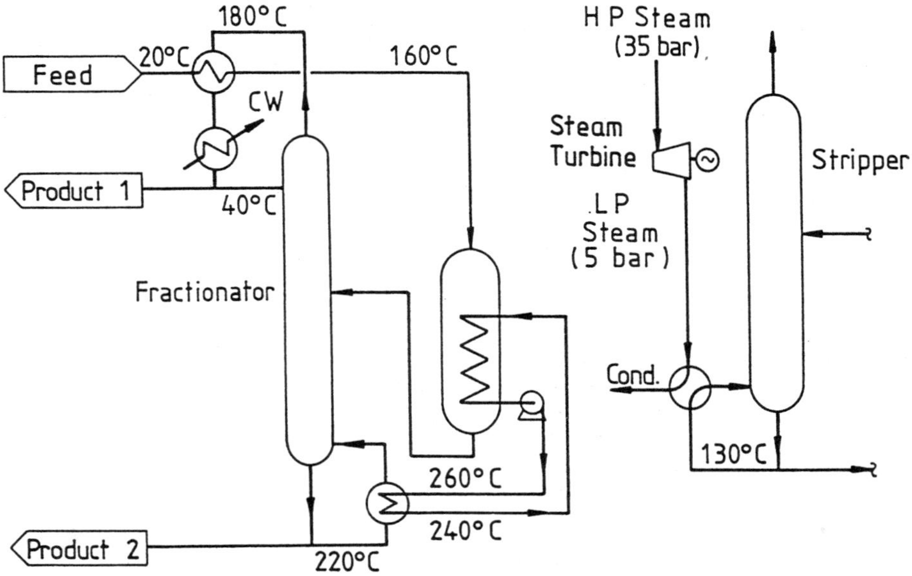

Figure 17

The improved CHP scheme

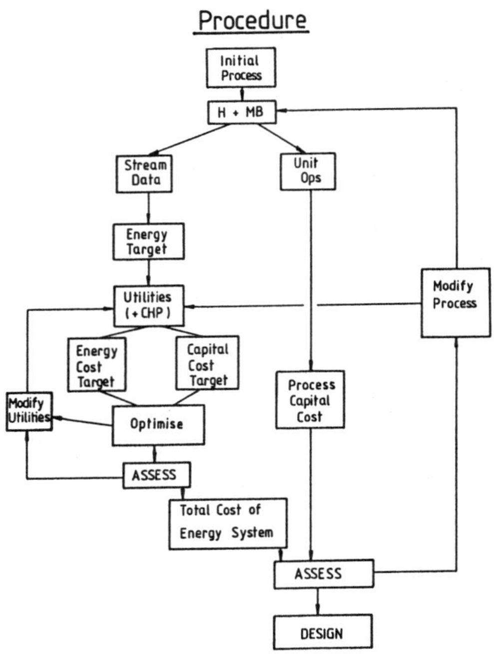

Figure 18

An overall approach for process design

PROCESS INTEGRATION IN A BENZOLE REFINERY

R. L. BARDSLEY
Staveley Chemicals

Staveley Chemicals

Staveley Chemicals operate 6 chemical process plants on a site at Chesterfield (near the centre of England) manufacturing

Sulphuric Acid
Chlorine/Caustic Soda
Benzene/Toluene/Cyclohexane
Dichloroaniline
Herbicides
Drugs (Paracetamol)

500 people are employed with production taking place round-the-clock 24 hrs per day, 7 days per week. The company turnover is over £40M per annum and the energy bill is in excess of £5M per annum.

Electricity for the chlorine cells is the largest component of this bill and its usage is under constant review. The second largest item is fossil fuels for the BTC plant and for steam to supplement the by-product steam from the Sulphuric Acid plant.

The Study

We made various ad-hoc attempts at energy saving in the 70's but only set up a formal Energy Saving Committee in 1984 to look at company wide projects. Contact with ETSU resulted in an invitation to Prof. Linnhoff to a lecture on Process Integration.

A consultancy project evolved to study the integration of the BTC plant, which was partly funded by ETSU and ran for 6 months in 1985. A secondary consideration of the project was to examine the energy balance across the whole site, and the BTC plant in relation to the site energy flows. Process Integration theory tells us we must do this.

The BTC Plant

This refines crude benzole, a by-product of coke oven gas manufacture, into pure benzene and toluene, and can also produce pure cyclohexane (B. T. & C.). The process is a high temperature catalytic continuous hydrogenation which is more correctly described as follows:-

1. Hydrogen is produced by steam reforming of hydrocarbons and hot 95% hydrogen is passed to the HOUDRY LITOL unit.
2. There the process involves firstly the fractionation of a C_7/C_9 liquid away from heavier residues (which are burnt as fuel) and vapourising it with the hot hydrogen. It then passes over a catalyst which reduces Sulphur compounds to Hydrogen Sulphide and dealkylates C_7, C_8 and C_9 aromatics to Benzene.
   The reaction products include a methane fuel gas which is used both for reformer feedstock and for heating; benzene, and unreacted toluene which are separated and purified by distillation.
3. Benzene can be further hydrogenated over another catalyst to make cyclohexane, so the final products are Benzene, Toluene, Cyclohexane plus Carbon Dioxide as a by-product of the reformer.
   In all, the process comprises over 40 streams of products being heated or cooled continuously - an ideal system for a Process Integration Study.

Results

The Process Integration study can be described as follows:

1. Firstly various operating scenarios were developed to allow for the wide variation of plant loadings experienced in practice, from 25% to 100% of rated load for the 3 separate units: Litol, reformer and cyclohexane. Pinch temperatures and heat requirements were assessed for each option.
   The pinch temperature of the whole process at 100% design was 114°C. With a minimum temperature gap of 20°C between the hot and cold streams, the minimum heat requirement of the process was estimated at 860 units relative to a current usage of 1000 units, that is, a reduction of 14%.
2. On further examination of the separate processes, most of the opportunities for saving occur within the Litol process, as the reformer and cyclohexane plants are well integrated.
3. The Litol pinch temperature was 140°C (Fig. I).
   The Litol heat requirement could be reduced in theory from 543 units (current usage) to 421 units, a reduction of 22%.
   Opportunities for actual process change (such as altering pressures, temperatures, etc.) were examined, but in the light of operability constraints it was felt these would have only a minimal impact on the heat requirement.
4. When the recovery of heat from the flue gas stream of the direct fired process heaters was considered as an additional stream, there was a further potential for heat recovery bringing the minimum heat requirement down to 299 units.
   (We had been aware for some time of this potential heat recovery opportunity but had encountered considerable difficulties in engineering a practical scheme).
   Considering the practicalities of gas/gas heat exchange and a temperature gap of 50°C, a target of 323 units was taken, that is, an overall reduction on current total heat demand of 40%.
5. The heat exchange network was studied (Fig. II): 2 examples of cross-pinch energy transfer plus 3 cases of utility heating below the pinch contributed to the present inefficiency. The next step was to redesign the network in 2 separate halves - firstly the network above the pinch and secondly the network below the pinch, to exploit the potential for savings.
   Various design options were proposed and rough costs estimated, e.g.
   i) Minimum temperature gap of 10°C (or even less for certain favourable exchanges).
   ii) Use of gas turbine exhaust for process heating.
   iii) Designs to maximise the saving of either
      a) steam,
      or, b) coke oven gas
      which are the main heating media.

Projects

Two of the schemes were subject to more detailed engineering and costing, as being the best options. These involved taking more heat from the main reactor effluent stream. These were

	Capital	Pay back
Scheme 1, 5 exchangers with temperature gap of 20°C	>£100,000	1 year
Scheme 2, 6 exchangers including the flue gas exchanger	>Scheme 1	1 year

The gas turbine scheme required substantially more capital and was eliminated at this stage. The chosen scheme maximised savings in process steam. Various smaller inefficiencies were also noted within the 3 separate process units of the BTC and remedial work incorporated within the overall project. Opportunities for the use of hot water at 80°C are still being studied.

The opportunity was then taken to examine the BTC process within the context of the total site operation. In practical terms this reduced quickly to a re-examination of the site steam system.

(In fact, if only the BTC plant had been built right next to the Sulphuric Acid plant, its heat demand would have dropped to zero).

The proposal was, however, to install at the Sulphuric Acid plant:

a) a back pressure turbine in place of pressure regulators;
b) a condensing turbine to deal with the steam excess in the summer months.

This scheme 3 had a total installed capacity of 3.8 MW and estimated payback of less than 4 years.

Our own engineers further modified scheme 2 to overcome specific operability constraints, reducing the heat exchangers by one unit and substituting a knock-out pot. We have proceeded to implement these schemes following Board approval for the capital as follows:-

Flue gas exchanger	July 86
4 Process exchangers, etc.	November 86
Steam turbines	March-June 87

Overall savings in total have been roughly in line with prediction and we expect savings eventually for schemes 2 and 3 to achieve £750,000 per annum - a significant reduction in the energy bill.

Fig I Litol Unit Composite Curves

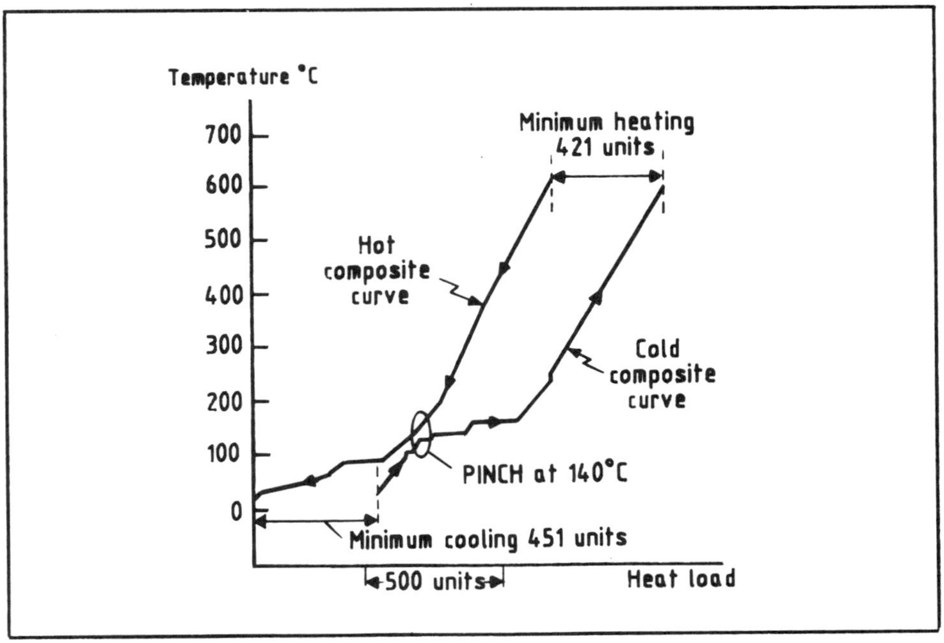

Fig II Litol Heat Network

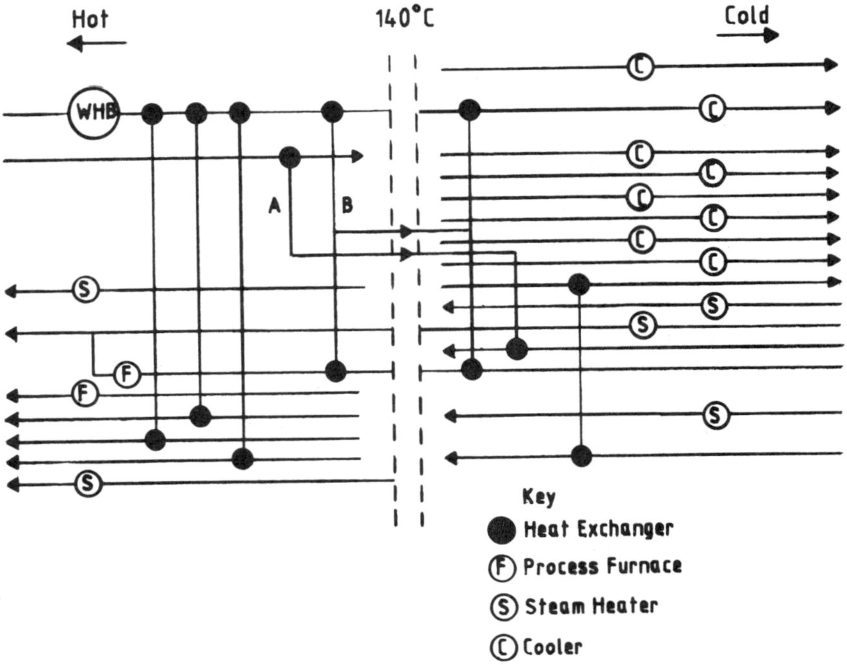

THE RESULTS OF A PROCESS INTEGRATION STUDY TO IMPROVE ENERGY EFFICIENCY AT A BRITISH BREWERY

R. MARSH C.Eng., M.I.Mech.E:
Chief Engineer and Energy Manager
Tetley Walker Limited, Warrington, England

1. INTRODUCTION (COMPANY BACKGROUND)

Tetley Walker is one of a group of six breweries which form the British sector of a company known as Allied Breweries Limited which, in turn, is the Beer Division of the parent company of Allied Lyons Limited which is probably the largest food and drinks company in Europe.

The breweries produce an extensive range of ales and lagers which are marketed on a nation wide basis, as well as overseas. Some of the more famous beers produced are "Long Life","Skol","Castlemaine XXXX" (under licence), "Tetley Bitter","John Bull" and Ind Coope Burton beers.

Tetley Walker is charged with the responsibility of producing mainly Tetley Bitter and Tetley Mild and Walker's Bitter and Walker's Mild in both cask-conditioned and brewery-conditioned qualities. The brewery does not produce lager. Beer is delivered by road to about 2,000 points of sale each receiving a delivery about once per week.

In total Tetley Walker employs about 2,500 people of which 700 operate in, or from, the Brewery.

The Brewery has a maximum production capacity of 33,000 hectolitres per week and covers an area of about 70,000 square metres.

2. THE PLANT AND ITS OPERATION

Because beer is a perishable product, and because of variable weekly order levels, the production targets are set on a week-to-week basis.

One 'Brewing cycle' consists of 900 HL. and in practice, weekly production targets can vary from as little as 16 'Brews' to as many as 30.

The higher the level of production the better are the energy ratios in terms of energy consumed per barrel produced (1.64 HL.).

Beer is brewed, fermented and conditioned on a continuous three-shift basis. Packaging into casks or kegs is carried out on a two-shift basis between 0600 and 2200 hrs., Monday to Friday, with major maintenance and housekeeping activities reserved for Saturdays.

3. MAIN SOURCES OF ENERGY CONSUMPTION

Diagram I is a schematic representation of our brewing, fermenting, conditioning and packaging processes.

Special points of interest to Linnhoff March Ltd., the Consultants who carried out the study, were :

3.1 The wort boilers which are heavy users of steam. More reclamation of heat from hot vapours generated by boiling of wort was thought to be possible.

3.2 The efficiency of the steam boiler plant and steam distribution network.

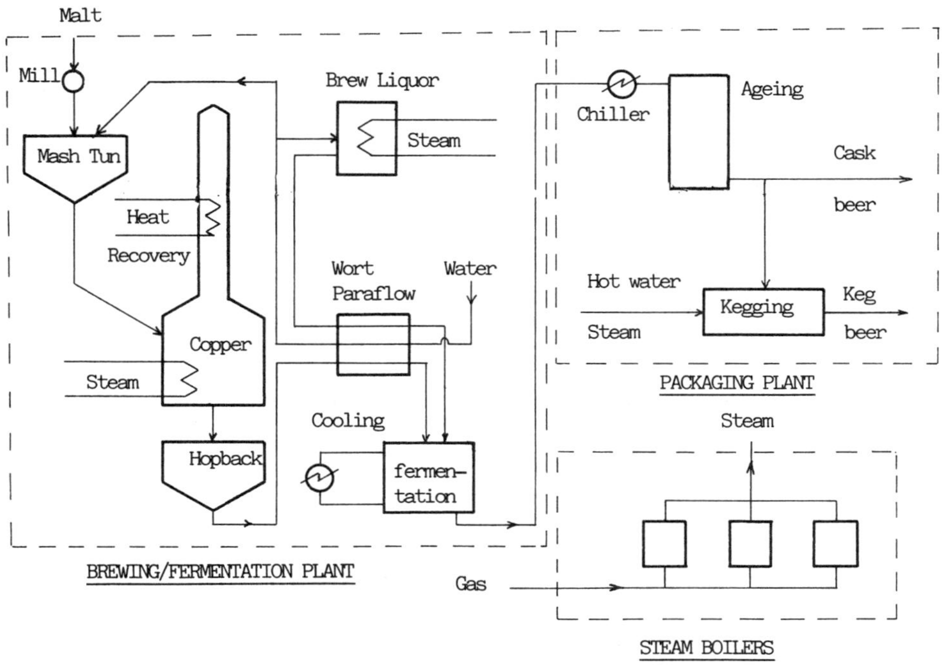

Diagram I. Brewery Production Cycle

3.3 The possibility of the use of direct gas fired heating to produce the large quantities of hot water for general cleaning duties (to replace steam as the heating medium).

3.4 The possible introduction of a combined heat and power plant to the Brewery to generate both steam and electricity, and thus reduce energy costs.

3.5 Space heating for offices, workshops and warehouses, i.e. the possible change from a steam distribution network to locally fired automatic gas boilers as the heat source.

TETLEY WALKER BREWERY ENERGY COSTS MAR. 1985 to FEB. 1986	£
FOSSIL FUELS:	460,000
ELECTRICITY:	423,000
TOTAL:	883,000

Diagram II

TETLEY WALKER LTD.

Process Integration Study

Maximum Consumption Allowance - Natural Gas

In 1973 max. allowable take was 3.4×10^8 MJ/yr.
(3.2×10^6 therms/yr.)

In 1986 max. allowable take reduced to: 1.9×10^8 MJ/yr.
(1.8×10^6 therms/yr.)

Reduction in max. take = 43.75% achieved by energy conservation initiatives prior to P.I. study commencing.

Diagram III.

4. BACKGROUND TO THE PROCESS INTEGRATION STUDY

In 1985 Tetley Walker Limited had implemented almost every energy conservation idea it could think of giving an acceptable return on capital invested.

A visit by Professor Linnhoff led the Company to believe that another approach to energy conservation did exist which could lead us to new ways of reducing energy costs i.e. Process Integration. The Company was able to obtain a Government grant towards the cost of such a study which was carried out during 1985.

The total cost of the study was £30,000.

5. THE RESULTS

TETLEY WALKER BREWERY
ENERGY DATA

1985/86 Unit Energy Consumption :-	=	135 MJ/Eq.HL.
Theoretical Company Target :-	=	100 MJ/Eq.HL.
Achievable Targets Identified by P.I. Study		
Adoption of Scheme "A"	=	111 MJ/Eq.HL.
Adoption of Scheme "B"	=	110 MJ/Eq.HL.
Adoption of Scheme "C"	=	119 MJ/Eq.HL.
Possible Improvement	=	18.5%

Diagram IV.

5.1 The British Brewing Industry, for reasons of energy usage comparison, has introduced a theoretical unit of production known as an "equivalent hectolitre" (EQ. HL). This system of comparision recognises that greater inputs of energy are needed for different types of beer and different forms of packaging the beer. For instance, Lager beer requires a greater input of energy to produce than ordinary beer and more energy is required per packaged H L when packed into bottles or cans, than in casks or kegs.

5.2 All British breweries are now able to convert their types of production and packaging to the national unit of EQ. HL. and hence they can compare their use of energy on a common basis.

5.3 At the time of the study the unit energy consumption at Tetley Walker was 135 MJ/EQ.HL. averaged over the previous year of production.

The P.I. study identified, for the first time to the Company, the absolute minimum levels of energy consumption it was possible to achieve for the prevailing production levels. This is an important factor for any industry, i.e. to know what is possible and what is not.

5.4 Three varying routes were identified which would lead to improved use of energy. Scheme "A" would result in a final energy input of 111 MJ./EQ.HL. whilst those for Schemes "B" or "C" would result in 110 and 119 MJ./EQ.HL. respectively (2).

Scheme "C" proposed the introduction of combined heat and power system, which, although giving the highest energy input per EQ. HL., would give the lowest energy cost per EQ. HL. by virtue of self generation of electricity which is the most expensive form of energy in Britain.

5.5 Schemes "A" and "B" were the routes chosen which included the actions shown in Diagram V :-

TETLEY WALKER BREWERY - WARRINGTON

Main projects to be implemented by the P.I. study :-

1. Improved heat reclamation from boiler flue gases.
2. Improved heat reclamation from hot wort.
3. Improved heat reclamation from copper vapour.
4. Improved heat reclamation from flash steam.
5. Improved reclamation of hot condensate.
6. Upgrading or addition of insulation.
7. Decentralisation of boilers for space heating systems.

Estimated total cost :- £170,000)
) Jan. 1986
Estimated payback period :- 2 years)

Diagram V.

Soon after the study was completed a large reduction in the price of natural gas (and oil) took place and "payback" periods would have been much better if this had not taken place, although all energy users were pleased that it did.

6. LESSONS FROM THE PROCESS INTEGRATION STUDY

6.1 Process Integration (P.I.) has a role in any industry in which energy is a substantial cost factor.

6.2 P.I. is best applied during the design stages for major plant renewals, or, even better, for completely new production centres.

6.3 When the P.I. study commenced Tetley Walker believed almost all avenues for improved use of energy were exhausted. The results of the study generated new thinking and a new energy conservation campaign.

6.4 The recommendations arising from the study could be pursued on a progressive basis i.e. as when the necessary capital became available.

(Most of the proposals contained in Schemes "A" and "B" have already been implemented and design work completed for the remainder).

6.5 Scheme "C", involving the use of combined heat and power, has not been adopted. The reasons for this are complex but it will be looked upon as a possible innovation when the present boiler-plant reaches the end of it's useful life, or should the cost of electrical energy reach unacceptable levels in the future.

REFERENCES

(1) LINNHOFF MARCH "Process Integration Study for Tetley Walker Limited, Warrington Brewery"

(2) ENERGY EFFICIENCY OFFICE (U.K.) "Cost Reductions on a Brewery Identified by a Process Integration Study at Tetley Walker Limited.

(3) ENERGY EFFICIENCY OFFICE (U.K.) "Tetley Walker Brewery Process Integration Study: A Summary.

SESSION III : NEW TECHNIQUES FOR LOW-TEMPERATURE HEAT RECOVERY

Harnessing heat pump and steam recompression technology to meet the needs of industry

Impact of new technologies on future heat exchanger design

Energy recovery by mechanical recompression of hydrocarbon vapour

Heat exchangers in plastic

Vapour compression in a brewery

Valorization of residual steam in brine evaporation

Overview of the European Community research and development actions of low temperature heat recovery

HARNESSING HEAT PUMP AND STEAM RECOMPRESSION TECHNOLOGY TO MEET THE NEEDS OF INDUSTRY

R GLUCKMAN MA CEng MIMechE MInstR
March Consulting Group, Windsor, UK

1. INTRODUCTION

Even though the concept of heat pumping was recognised more than one hundred years ago, the technology has a long way to go before it reaches maturity. Only a small proportion of the energy saving potential of heat pump applications in European industry has been achieved. A major reason for the lack of installations is related to fuel costs; even after the fuel crises of the 1970's energy prices were not high enough to encourage the technology of heat pumps to be fully developed. However, there has been another cause of disinterest - heat pumps have in many cases proved unreliable and have not met their design performance.

In 1987 we are at an interesting watershed in the history of the heat pump. We have 10 years of vigorous design activity behind us. More than one thousand industrial systems (over 100 kW and up to several MW) have been installed in Europe, North America and Japan. These heat pumps can be thought of as "first generation" systems. Although in some cases they were unsuccessful in economic and engineering terms, many useful lessons can be learned from this experience. If we can benefit from previous mistakes (and successes!) then it is possible to envisage a "second generation" of heat pumps that will achieve high levels of performance and reliability. If, on the other hand, we fail to use the existing base of knowledge and experience then the industrial heat pump market in Europe has little chance of development.

In this paper the potential for industrial heat pumps in Europe is investigated and assessed. First brief reviews of industrial heat pump technologies and markets are made. Then a summary of useful technical design guidelines is presented.

2. INDUSTRIAL HEAT PUMP TECHNOLOGY

The industrial heat pump encompasses a large range of system designs and variants. It is difficult to rigorously define exactly what is meant by an "industrial" heat pump. In the author's view two basis criteria must be met:

* the heat pump must involve industrial process heat (either as the heat source or the heat user or as both)
* the heat pump must be of industrial scale ie a minimum size of 100 kW heat output.

Some of the largest heat pumps in the world are used for district heating systems. With heat outputs of over 10 MW these are clearly of industrial scale. However, unless they involve an industrial waste heat source they cannot be truly considered as industrial heat pumps.

Industrial heat pumps fall into three main technology groups:

* closed cycle compression
* open cycle compression
* absorption cycle

Each of these basic types is described including reference to important system variants, application areas and technical limitations.

2.1 <u>Close Cycle Compression</u>

The most familiar type of heat pump used in industry is the closed cycle compression system. It is shown in its simplest form in Figure I. A closed circuit refrigerant loop exchanges heat with the waste heat source and the heat user.

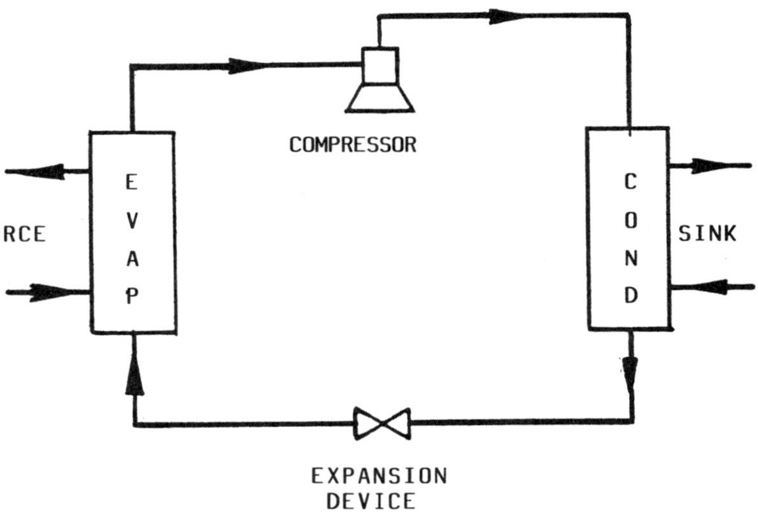

FIGURE I SINGLE STAGE HEAT PUMP

The system comprises four essential components:

* An evaporator in which waste heat is absorbed into a boiling refrigerant.
* A compressor which raises the pressure, and hence temperature, of the refrigerant.
* A condenser in which the absorbed energy and compressor shaft power are released to the heat user.
* An expansion device in which the condensed refrigerant liquid changes from a high temperature liquid to a low pressure (and temperature) mixture of liquid and vapour.

There are many cycle variants and it is unlikely that an industrial system will be most efficient (or most cost effective) in the form shown in Figure I. Important options that should always be considered at the design stage include:

a. Use of a refrigerant subcooler to preheat the heat user stream with liquid refrigerant from the condenser.

b. Use of a refrigerant desuperheater.

c. Cascaded heat pumps (the use of several small heat pumps operating in series instead of one large unit).
d. Engine driven compressors with waste heat recovery.

The choice of the best cycle is very site specific and depends on technical factors (such as heating temperature range of the heat user, the overall temperature lift between source and user) and economic factors (such a fuel prices and annual operating hours).

Similarly, the components are available in a wide range of types. Small systems (less than 250 kW) usually use reciprocating compressors. Larger plants either have screw or centrifugal compressors. Evaporators and condensers are usually shell and tube for liquid heat sources/users or finned coils for gaseous sources/users. Refrigerant fluids are usually of the halocarbon ("Freon") type although there is no reason to prevent other fluids being used in appropriate circumstances (eg ammonia). R22 is suitable up to user temperatures of around 50°C. R12 is used up to 80°C and R114 or R500 above this figure.

The closed cycle compression heat pump can be used for a wide range of applications. Sources can include streams of water, air, steam or any other liquid or vapour. Careful selection of heat exchanger materials and design can allow dirty or corrosive streams to be used. The main technical limitation is related to maximum temperature. Above condensing temperatures of 120 to 140°C most applicable refrigerants suffer thermal degradation, particularly in the presence of lubricating oil; this prevents their usage above these temperatures. In economic terms the application of closed cycle systems is also limited by temperature lift. If the temperature lift is too large then the heat pump performance will not be good enough to compete with fossil fuelled heating systems. Maximum acceptable lift is very site specific but is in the region of 50 deg C for simple cycles and 80 deg C for more efficient variants.

2.2 Open Cycle Compression

The open cycle system does not have a special refrigerant fluid as described above; it uses the process steam itself as the fluid to be compressed. The open cycle system is often referred to as Mechanical Vapour Recompression (MVR) or, if water is the process stream, steam recompression. A typical configuration is shown in Figure II. This figure illustrates MVR being used in its commonest application ie evaporation or concentration. The solution to be concentrated is boiled at atmospheric pressure, giving off water vapour at 100°C. This vapour is compressed and supplied at higher and temperature (say 2 bar(a), 120°C) to the boiling heat exchanger.

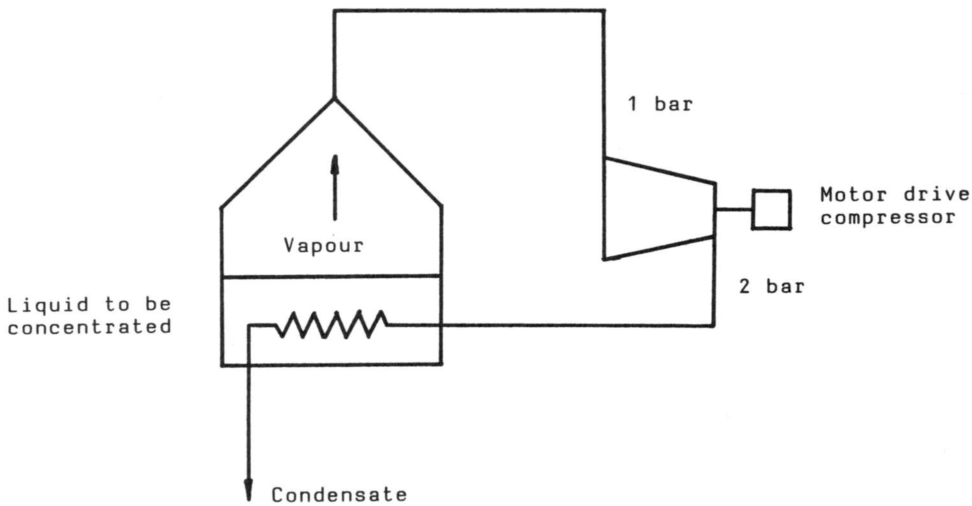

FIGURE II OPEN CYCLE HEAT PUMP

In essence the MVR system is simply a compressor fitted between the process vapour exhaust and the boiling heat exchanger. As such, it has no major design variants. However it is not necessarily easy to modify an existing evaporator to act as an open cycle heat pump. In particular it is very important to minimise the heat pump compression ratio if good efficiency is to be obtained. This usually means using a very large heat exchanger in the evaporator. For example, a conventional atmospheric pressure evaporator may use steam at 5 bar(a) which is equivalent to 152°C. This pressure is much too high for efficient heat pumping. A much larger heat exchanger must be fitted so that steam at 2 to 3 bar(a) can be used to boil the solution.

The main option in choice of components relates to the compressor. The most commonly used types are centrifugal, screw and other rotary machines (eg rotary vane, Roots blower).

Applications are much more restricted than for closed cycle machines. The most promising areas of use are for evaporation, concentration and distillation. Some drying processes are being adapted for MVR operation. Although there are relatively few different applications, MVR is a very important technology. Very high efficiencies can be obtained; this means that significant market penetration will be easier than for other heat pump types.

2.3 Absorption Cycle Heat Pumps

Both closed and open cycle systems require a compressor. This is an expensive and complex piece of equipment. Absorption cycle systems do not use a compressor and use heat as the energy input in place of shaft power. As shown in Figure III, the system still has a refrigerant condenser, expansion valve and evaporator. However the compressor is replaced by an absorber, generator and liquid pump. Low pressure refrigerant vapour is passed into the absorber where it is dissolved in water. The solution is pumped to the generator where it is heated. Refrigerant vapour is released and passed to the condenser; dilute absorbent is passed back to the absorber.

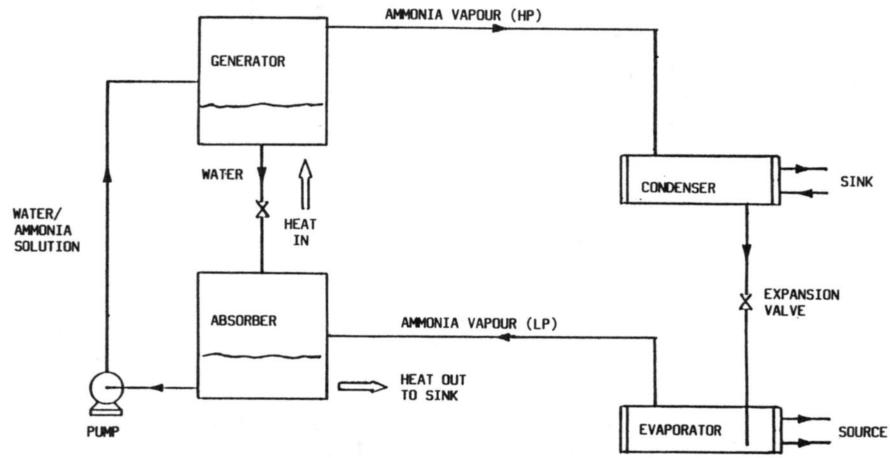

FIGURE III AMMONIA/WATER ABSORPTION CYCLE

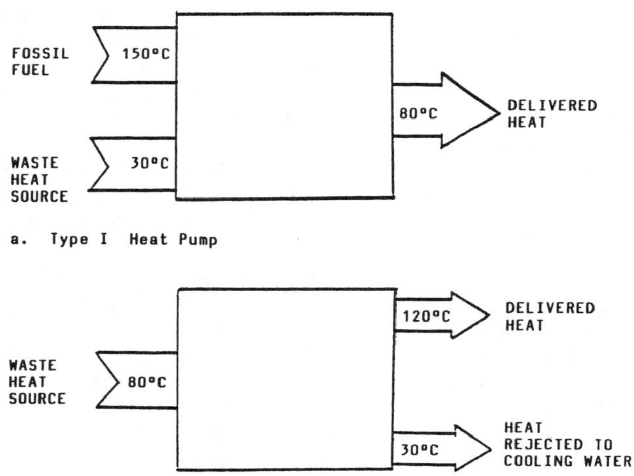

a. Type I Heat Pump

b. Type II Heat transformer

FIGURE IV ABSORPTION SYSTEMS

There are two basis variants, the type I absorption heat pump and the type II heat transformer. (Note, in the USA these variants are respectively known as heat amplifiers and temperature amplifiers). These variants are shown in Figure IV. The type I system has two heat inputs (waste heat at low temperature and fossil fuel derived heat at high temperature) and a single heat output at a medium temperature level. The type II system only has waste heat as input. A proportion of this heat is raised in temperature; the remainder is rejected at a low temperature.

Much research is taking place to identify refrigerant/absorber pairs. The most commonly used heat pump pairing is Lithium Bromide/water (with LiBr as absorber and water as the refrigerant). Most of the industrial systems are very large (heat output over 1 MW) and of Japanese origin, although European manufacturers are also active in the field.

Applications are quite varied. The main technical limitation is temperature lift which is restricted to 50 deg C for single stage systems using LiBr/water. Top temperatures are also restricted to 120-140°C because of the onset of corrosion and crystallisation problems.

3. THE EUROPEAN INDUSTRIAL HEAT PUMP MARKET

There are now over 800 large industrial heat pumps (>100 kW) installed in the European Community. These include a wide range of sizes (up to several MW) and designs. The widespread application of industrial heat pumps followed the 1972/3 oil price rise. The build up of installations is shown in Figure V. From this pattern we can see the decline in demand that has coincided with the 1985/6 fall in oil price.

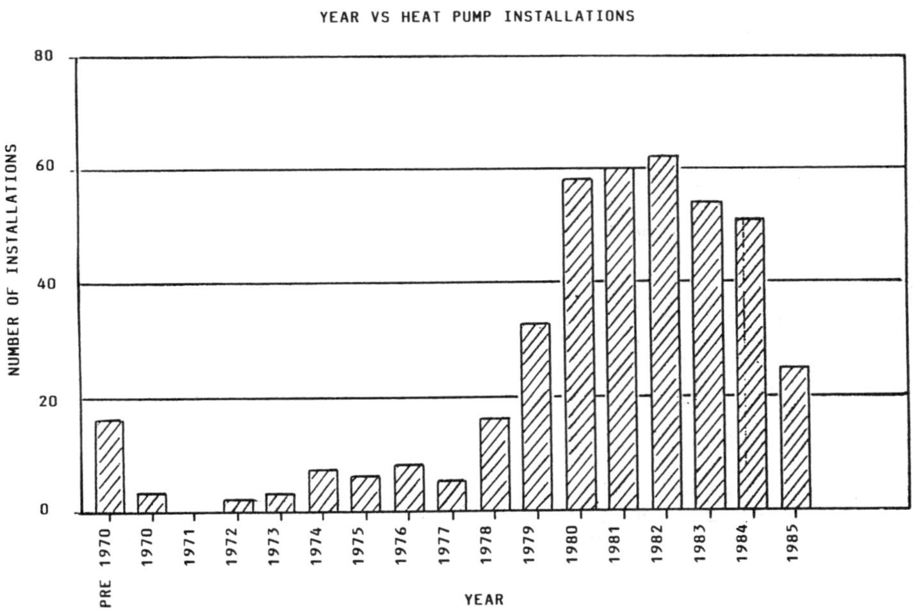

FIGURE V INDUSTRIAL HEAT PUMP INSTALLATIONS IN EUROPE

The commonest systems are closed or open cycle compression. To date only a few absorption systems have been used in Europe. The open cycle system is emerging as the most successful category of industrial heat pump. Very high coefficients of performance (COP, the ratio of heat output to energy input) can be obtained with open cycle plant. This makes MVR very competitive.

There are significant regional variations in the adoption of industrial heat pumps. France and Germany clearly lead the EEC. In France electricity prices are very low,which favours heat pumps. In Germany the support of Government has encouraged the use of heat pumps in industrial as well as commercial and domestic sectors.

The industrial sectors with most heat pump applications are the "low temperature" process industries including chemicals, food, drinks, textiles and paper.

4. GUIDELINES FOR GOOD HEAT PUMP DESIGN

Analysis of the performance of existing "first generation" heat pumps has shown that a number of design faults are very common. By avoiding these problems it will be possible to design new heat pumps with much higher standards of reliability and techno-economic success. Many of the problems relate to four general reasons:

a. Use of Refrigeration Rules of Thumb

It was generally assumed that heat pumps could be built with exactly the same design parameters as refrigeration plant. In fact this is not the case. Great care must be taken in extrapolating refrigeration data.

b. Use of Fossil Fuel Rules of Thumb

In a similar way it is dangerous to design a heat pump process heater in the same way that a fossil fired heat would be designed.

c. General Lack of Detailed Design

Many heat pumps were installed without enough initial design work. The heat pump is typical of many post war technologies that has suffered from an excess of enthusiasm and expectation and a great lack of attention to detail.

d. Poor Manufacturing Standards

In many cases too many economies were made to keep capital costs low and heat pumps consequently suffered from unnecessarily poor reliability. This is particularly true of site installation work.

4.1 System Design

a. System Sizing

The most common problem with first generation heat pumps is oversizing. Systems were built to meet peak process heat demand plus a contingency allowance (in the same way boilers are design). This is wrong for two reasons:

- heat pumps do not generally perform well at part load.
- heat pumps are relatively expensive and must run for a large number of hours per year to achieve good economics.

An important rule is to make heat pumps large enough only to provide base heating load. In a number of cases heat pumps achieve good thermal performance in terms of COP and heating duty but have had very poor payback periods because of the low number of full load running hours.

b. Correct Choice of Thermodynamic Cycle

Many system designers think of heat pumps only in terms of simple four component cycles. This is convenient for manufacturers who can provide standard packages in a range of sizes. However, the opportunity to improve system performance is often lost. Incorporation of liquid subcoolers, desuperheaters, horizontal cascade cycles and two stage systems can often give higher COP with little or no extra capital cost.

c. Incorporation of Passive Heat Recovery

Passive heat recovery is always cheaper in terms of capital cost than a heat pump and, of course, has no appreciable use of energy for operation. It is vital that all opportunities for passive heat recovery are used before heat pumps are considered.

d. Process Integration

A powerful new technology now exists to identify the correct way to apply heat recovery to industrial processes. It is called Process Integration or Pinch Technology and gives very important rules about the application to heat pumps.

It is not possible to properly explain Process Integration (PI) in this paper but it is well covered in the literature. A PI analysis identifies a unique temperature in any industrial process called the "Pinch Temperature". Knowing the pinch point can help the process designer in may ways. For heat pumps the important rule is:

* Heat pumps should only be used "across" the pinch, ie the heat source should be below pinch temperature and the heat user should be above (Figure VI).

4.2 Component Design

a. Evaporators

Poor evaporator design has led to many problems in first generation heat pumps. The faults lead to poor heat transfer coefficients and a loss of both COP and thermal capacity through low evaporating temperatures. Particular problems include refrigerant distribution, excessive superheat, effects of oil, evaporator fouling and corrosion. It is recommended that flooded evaporator circuits are used in preference to direct expansion systems.

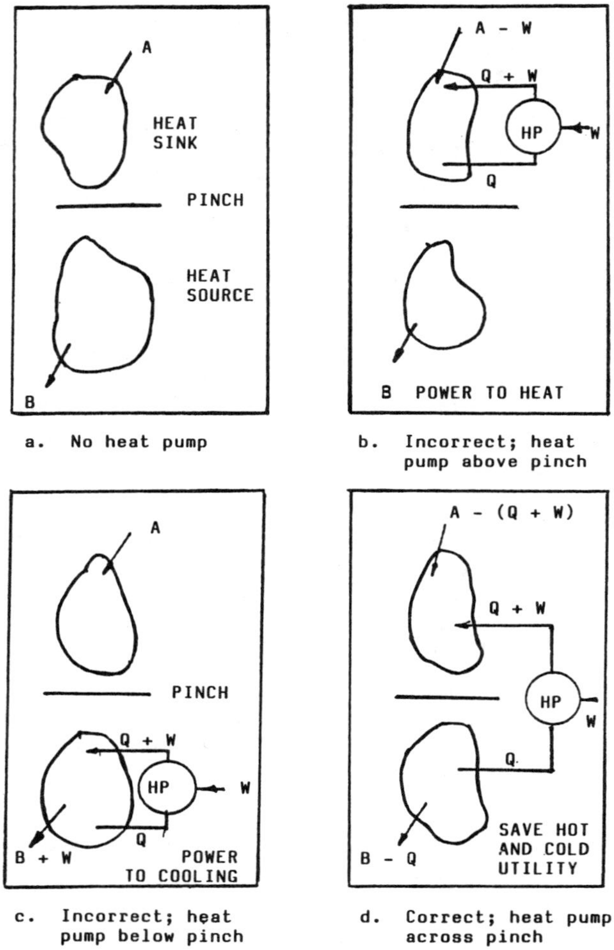

FIGURE VI PLACEMENT OF HEAT PUMPS

b. <u>Compressors</u>

The compressor has seen major improvements during the last decade for heat pump applications. At first it was believed that completely standard refrigeration compressors could be used. This has been found to be wrong except for extremely robust machines that were previously overdesigned. Great care must be taken using the cheaper machines that are designed for packaged chillers and air conditioning plant. Some common faults have included valve and bearing failure; shaft seal leakage; poor efficiency; failure of unloading gear and ancillaries; motor burnouts.

c. <u>Condensers</u>

Of the three main components of heat pumps the condenser has caused the fewest problems. The main cause for concern has been fouling or corrosion on the heat user side and correct piping of multi-condenser systems.

d. Engine Driven Systems

Engine driven systems are often worth considering because they improve the overall energy saving potential (particularly for applications with high temperature lift). However, the engine does lead to a lot of extra design and maintenance considerations.

One of the classic lessons learned (and, unfortunately, relearned) during the last 10 years has been the simple rule about engine combustion air intakes. These MUST be ducted from outside the engine room in a position where refrigerant cannot be ingested into the engine. The consequences of halocarbons entering an engine combustion chamber are dramatic. The refrigerant is broken down by high temperatures into highly corrosive compounds of fluorine and chlorine. Severe engine damage is inevitable.

Other engine related problems have included engine/compressor vibration, exhaust heat exchanger design and use of lubricating oil.

e. Refrigerant Leakage

Many heat pumps have suffered with refrigerant leakage problems. In general the cause has been poor manufacturing standards, lack of checking and lack of attention to detail.

f. MVR Compressor Efficiency

One of the biggest problems of MVR systems has been poor compressor isentropic efficiency. Many plants had not achieved the efficiency claimed by manufacturers. A purchaser should obtain guaranteed efficiency data before ordering plant and this should be verified on commissioning.

g. Absorption Plant Corrosion

Recently one large heat transformer in a European chemical plant has had tremendous corrosion problems. The problems are believed to be related to the relatively high temperature of this plant (140°C) which increases the corrosivity of the LiBr. It should be noted that the cause of this problem has not yet been confirmed. Another possible cause is air inleakage which must never be allowed on absorption systems.

5. THE FUTURE FOR INDUSTRIAL HEAT PUMPS IN EUROPE

The potential for harnessing heat pump technologies in European industry is basically related to the overall level of fuel prices and to the ratio of fossil fuel prices to electricity. Given the low level of oil prices in 1986 (below $15/barrel) it would not be possible for heat pumps to establish a significant market. At the price levels of the early 1980's ($25-30/barrel) industrial heat pumps began to develop quite quickly (Figure V). Much practical experience has been gained from these installations. This means that if the oil prices rise back to $30/barrel a new generation of more efficient, reliable and well designed systems can be envisaged. Such systems will have greatest success in the MVR markets eg evaporation and distillation.

While oil prices remain low it is important that efforts are made to learn more about first generation heat pump experience. It is unlikely that manufacturers will invest much money in the field because of the lack of business. It is the role of European governments and the EEC to ensure that money is well invested into heat pump technology. Many Governments have taken a very short term view and have stopped supporting heat pumps at the very time when their help is most needed. Others are misdirecting their efforts into long term R & D in fields such as high temperature heat pumps and use of non-azeotropic mixtures. Whilst these topics are interesting and useful they will never make a significant change to the size of the industrial heat pump market. The important activity is to refine "conventional" technology so that European manufacturers are ready to meet the needs of the 1990's.

References

1. Gluckman, R. Heat pump cycles are their engineering. Part of booked entitled "Heat Pump for Buildings", ed Sherrat, published Hutchinson (1984).
2. Perry E J. Drying by cascading heat pumps, Proc 3rd Int Conf, Future Energy Concepts, IEE Conf Publ 192, IEE, London (1981).
3. Linnhoff B et al. User guide on process integration for the efficient use of energy, IChemE, London
4. Linnhoff B and Vredeveld D R. Pinch technology has come of age, Chemicals Engineering Progress (July 1985).
5. McMullen, J T, Hughes D W and Morgan R. Influence of Lubricating oil on heat pump performance, European Commission Energy RRD Programme, Contract EEA-4-028-GB.

IMPACT OF NEW TECHNOLOGIES ON FUTURE HEAT EXCHANGER DESIGN

D.A. REAY
David Reay & Associates,
PO Box 25, Whitley Bay, UK.

SUMMARY.

Heat recovery technology has featured strongly in most major industrial energy conservation programmes, and significant penetration of heat exchange equipment into processes, for the express purpose of energy efficiency, has been achieved. Barriers, both technical and economic, still exist however, preventing more widespread adoption of heat exchangers. Problems associated with fouling and corrosion remain, and the temptation to adopt more sophisticated energy recovery methods such as organic Rankine cycle machines has led to some over-ambitious installations with dubious economic benefits.

The prospects for cost-effective low temperature heat recovery are improving due to a combination of developments, including the novel use of materials and new heat exchanger concepts. A number of peripheral aids, in particular improved design procedures, including process integration, and the use of artificial intelligence techniques, can help users select appropriate state-of-the-art equipment.

In this paper, the role of techniques for enhancing heat (and mass) transfer will be discussed. Process intensification, perhaps normally associated with compact chemical plant unit operations, has, the author believes, an important future role to play in enabling heat exchanger size and costs to be reduced. Enhancement of transport processes, by techniques which are largely established in other heat transfer areas, or in other technologies, is a necessary step in achieving compact systems.

1. INTRODUCTION.

There was a temptation to give this paper a supplementary title of 'Small is Beautiful'. The tendency towards miniaturization of engineering systems, within which the term 'process intensification' may be categorised, brings to mind efforts at Nijmegen University some years ago, with which I was associated. These were to develop and substantiate the hypothesis that the 'resting' eccrine sweat gland acts like a heat pipe (1). The heat pipe, of course, is an element of many gas-gas heat recovery systems but the principal role of its biological equivalent is not heat conservation, but water retention!

I was pleased to read recently that mechanical engineering had, at least in part, caught up with biological engineering capabilities, in that 'micro heat pipes' can be made with lengths of a few cms, and diameters of 10-500 microns (2). The application cited is for cooling of microelectronics and other small devices. Spacecraft applications of heat pipes provide an insight into the degree of 'intensification' demanded by energy dissipation needs on advanced orbiters. These are evident from the list below.

Technology.	Capacity (kW-m).
Axial groove heat pipe.	0.25
Monogroove heat pipe.	25
Capillary-pumped plate.	250
Pumped two-phase loop.	25,000

Of the above, only the first is routinely used, while the others are under development. In the concepts I will be discussing here, the individual enhancements may reach one order of magnitude, although cumulative benefits have yet to be analysed. (Note the dimensions of the capacity - this relates to the distance over which a given quantity of heat can be transported, less denanding in terrestrial applications, of course).

The word 'miniaturization' is only quantifiable when one has an insight into what is being miniaturized. A compact chemical plant may still occupy several hundred cubic meters, while a miniature heat exchanger can have a surface area of a few hundred microns.

Miniaturization of mechanical systems has in part been helped by materials technology, and associated fabrication breakthroughs. For example, micromachining of silicon makes it possible to build engineering components, including mechanisms, almost as small as microelectronic components. These have included valves, nozzles and heat sinks. A heat sink developed to cool a silicon chip has been reported as having sub-surface channels of 300 x 50 microns, at a pitch of 100 microns. The cooling capacity using forced circulation of water is 1 kW/sq.cm; (in comparison, forced air systems conventionally cope with 2 W/sq.cm). This of course is a form of enhanced heat transfer which could be highly relevant to heat exchangers for heat recovery.

There are many other forms which are discussed in greater detail below, but lest I be immodest enough to claim originality of thought in this paper, it is salutary to refer to heat transfer texts such as that by Dr. Hryniszak on gas turbine heat exchangers (3) published in 1958, ie some 30 years ago. The main trends in the development of (gas turbine) heat exchangers were:

'Increasing the effectiveness...Reducing the size....Improving the design...and...Improving the cleaning facilities.' He also made reference to rotation as an enhancement technique for gas-side heat transfer. It is not the concepts which have changed over the years, more the materials and fabrication technologies which now assist us to realise some of these concepts.

2. PROCESS INTENSIFICATION.

Miniaturization, per se, does not solve any problems in fluid dynamics or heat and mass transfer, and it is the use of the term 'process intensification', in particular exemplified by the work of the UK chemicals company ICI, wherein lies the key to what could be a potentially valuable direction of research, development and, of course, demonstration. (In the UK, for example, government support for a demonstration of a heat exchanger with volumetric heat transfer coefficients 5 times those of conventional plate liquid-liquid heat exchangers is being explored).

The particular system just mentioned uses the heat transfer benefits of laminar flow in small channels (with proper respect for the fouling potential of such designs). Other intensification concepts exist. The National Engineering Laboratory in the UK pioneered enhancement of heat transfer using a liquid phase in air cooled heat exchangers (4). It was shown in the 1960's that the introduction of an atomised liquid brings about an order of magnitude increase in outside heat transfer coefficient on tubes of air coolers.

Electrical enhancement techniques have been successfully demonstrated on boiling refrigerants and other heat and mass transfer processes. These form an element of the heat exchanger work in the Japanese 'Super Heat Pump' programme. More recently the concept of electroacoustics, effectively combining electric and ultrasonic fields, has been applied to processes. The enhancement of chemical reactions using acoustics alone - sonochemistry - may also be included as being of interest to heat transfer engineers. Jet impingement is used to enhance convection features in the European Commission Non-Nuclear Energy R & D Programme.

Possibly one of the most interesting concepts, and the most difficult to successfully engineer, is the use of rotation as an enhancement method. Most listings of the relative merits of heat exchangers for heat recovery include as a 'plus' point - no moving parts. The emphasis given to this aspect is not of course particularly strong; after all, rotating regenerators are highly efficient gas-gas heat recovery units and the incorporation of pumps or fans is not regarded as a major hurdle.

The suggestion that whole systems, such as absorption cycle heat pumps or distillation plants and their associated reboilers, be rotated to improve effectiveness can, however, create a degree of scepticism. Nevertheless, a study of the recent patent literature reveals similar proposals from ICI, (and of course the Higee distillation column is now being marketed). A study of NASA projects undertaken in the 1960's shows how rotating boilers can be much more effective in terms of steam-raising capacity/unit volume than more conventional systems.

Without wishing to dwell on a concept which I have already used to illustrate another point, the rotating heat pipe has been demonstrated as one method for overcoming the heat transfer limitations of its static counterpart.

3. LAMINAR FLOW HEAT EXCHANGERS.

The attainment of high heat transfer performance using laminar flow has been mentioned above. The benefits resulting from the use of the 'Printed Circuit Heat Exchanger', (PCHE) as components of the evaporator and condenser of a packaged water chiller may be illustrated with reference to Fig. 1. The heat exchangers are less than 30% of the size of conventional shell and tube heat exchangers, but do not have the internal pressure limitations commonly associated with plate heat exchangers. Additionally, they are not restricted in their application by gasket material considerations.

The attractiveness of such heat exchangers in permitting major reductions in the volume of plant, and inevitably in the capital and installation costs, is important in many applications. The retrofitting of plant, particularly for heat recovery duties, is sometimes difficult because of space restrictions. The PCHE, which can also be used as a gas-liquid heat exchanger, has a degree of compactness which should give it and its derivatives an assured future.

Cross and Ramshaw (5), who are arguably the main inspiration behind the process intensification concept, have assessed the performance of the PCHE and similar heat exchangers. They make a particularly valid point concerning the retrofitting of compact plant, pointing out that if the maximum benefits are to arise out of the use of process intensification techniques, a reassessment of plant arrangement is necessary. Ideally, process intensification philosophies should be simultaneously applied to all the major items of the process plant being investigated. I will return to this theme later.

In determining the behaviour of the PCHE's, experiments were carried out at ICI on two basic forms of the unit. In one case copper plates were etched, forming channels to a depth of 0.3 mm, and having a similar width. A second heat exchanger was made up using

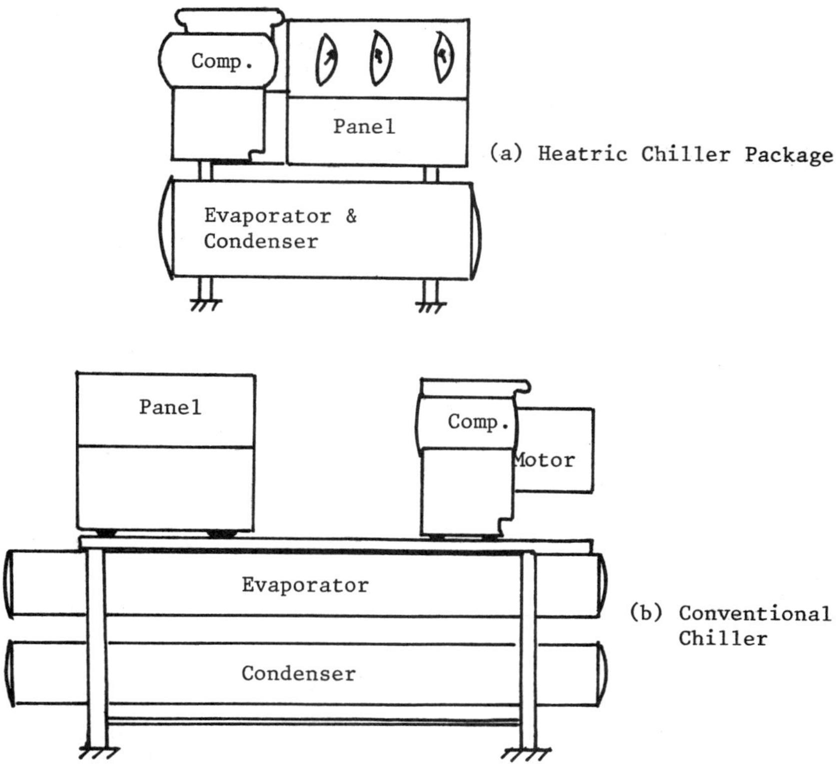

Fig. 1. The Heatric Chiller Uses a 'Printed Circuit Heat Exchanger'

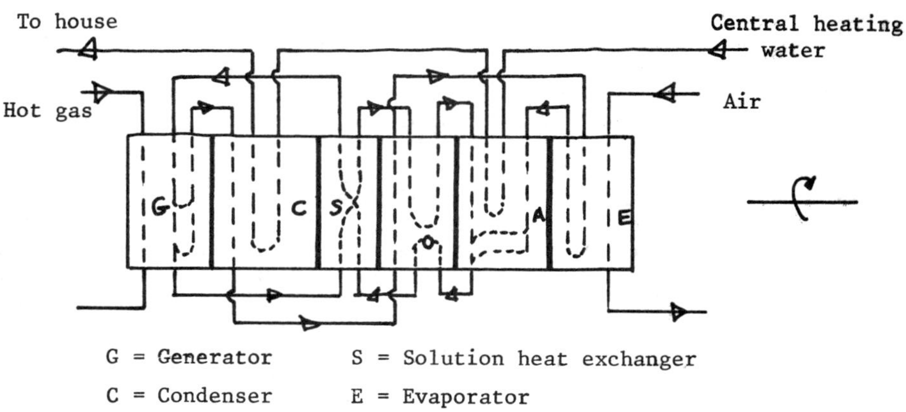

Fig. 2. The Rotating Absorption Cycle Heat Pump Patented by ICI plc.

corrugated titanium plates having channels which formed to give a hydraulic diameter of 1.19 mm.

Both heat transfer and fouling data were obtained in a series of experiments on purpose-built rigs. With regard to the former, the etched matrix, although conceded to be less than perfect in construction, had a volumetric heat transfer performance equivalent to 7 MW/cubic m.K with a water velocity of 0.18 m/s. The titanium plate unit, with a velocity substantially higher at 1 m/s, had a volumetric heat transfer coefficient of 7.3 MW/cubic m.K. Corresponding values for the 'equivalent' shell and tube and plate heat exchangers were listed as 0.21 MW/cubic m.K and 1.25 MW/cubic m.K respectively.

With regard to fouling, a silt having a particle size ranging from 2 to 90 microns, with a mean of 12 microns, was used. Trials were conducted on single plates, over periods of only a few hours. Nevertheless, encouraging data were obtained by reverse pulsing the flow to remove built-up deposits.

4. ROTATION.

The use of rotation to aid separation processes is not new. Centrifugal separators are at a highly developed state and are energy efficient. However, the use of rotation to enhance heat and mass transfer is less well-known, although practiced to a limited extent.

I have already made reference to the rotating heat pipe, where a major benefit accrues to the improvement in condenser performance achieved, and the complementary rapid return of liquid to the evaporator. The Higee distillation unit (perhaps column would be a misleading word because of the degree of compaction achieved) is a working example of a mechanical process with lower energy demands replacing a thermal system.

The absorption cycle heat pump has often been likened to a collection of heat exchangers, and extending the theme of the rotating boiler mentioned earlier, the application of the process intensification philosophy to heat pumps could have important implications. The subject of US Patent 4553408 (6) is a rotating absorption cycle heat pump which has already been the subject of experimental studies. The assignee is ICI.

It is not my intention to dwell on heat pump technology in this paper; suffice to say that the evaporator and condenser have enhanced performance due to rotation sufficient to produce forces in the range 100-600g, as do the generator (the 'rotating boiler') and the absorber. In the latter case benefits result to both heat and mass transfer. The unit, which would be much smaller than current absorption cycle heat pumps, is illustrated in conceptual form in Fig. 2. Interestingly, the patent suggests a working fluid pair of R124 (not a

CFC) and pentaoxapentadecane.

The reader is left to ponder the fine engineering detail, and to extend the thinking to vapour compression cycle systems.

5. CATALYSIS.

Approximately ten years ago I investigated the forms of heat exchanger which might benefit from a catalytic coating on one or more of the surfaces. The study was prompted by work on catalytic combustion in gas turbines in the laboratory where I was working at the time, and by a study of catalytic incineration techniques. Catalytic combustion is increasingly used to remove hydrocarbons and other combustible pollutants from exhaust streams prior to discharge to the atmosphere. A feature common to many of these plants is the need to preheat the process stream prior to catalytic treatment. This makes such a configuration an obvious candidate for heat recovery, and systems incorporating waste heat boilers and other heat exchanger types are numerous in the textile and chemical industries.

It would seem sensible to combine the role of catalytic combustion and heat recovery, instead of, as is the common case, having two sometimes large items of plant, as shown in Fig. 3. A waste heat boiler with finned tubes externally coated with an appropriate catalyst would appear a logical solution. Theoretical treatments of the enhancement in heat transfer arising out of surface combustion on finned tubes naturally suggest major improvements in heat transfer coefficients, and compact shell and tube catalytic heat exchangers have been produced in the past for aerospace use.

The experiments confirmed the possibility of catalytic combustion on a coated heat exchanger using a propane-air mixture, with the reaction self-ignition occurring at about 250 deg C. The work highlighted the need for close control of surface conditions, as ebullient cooling by water on the insides of the tubes led to quenching of the catalytic combustion. Nevertheless, as a concept for intensification of heat transfer where combustion products need to be separated from the sink medium, it is worthy of further study.

6. ELECTRICAL ENHANCEMENT OF BOILING.

The electrical enhancement of evaporation and condensation, sometimes called electrohydrodynamic enhancement, has been studied extensively in the laboratory, but only recently, in common with a number of other electrical enhancement techniques, has it received serious attention from equipment designers. In the UK, work at Imperial College and the City University, London, has established an experimental and theoretical basis for the determination of boiling and condensing

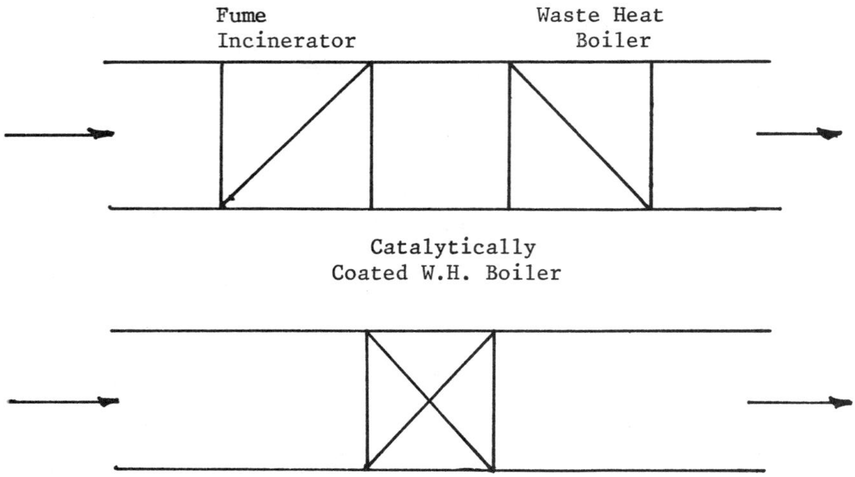

Fig. 3. Compact processing by Combining Fume Incineration and heat recovery.

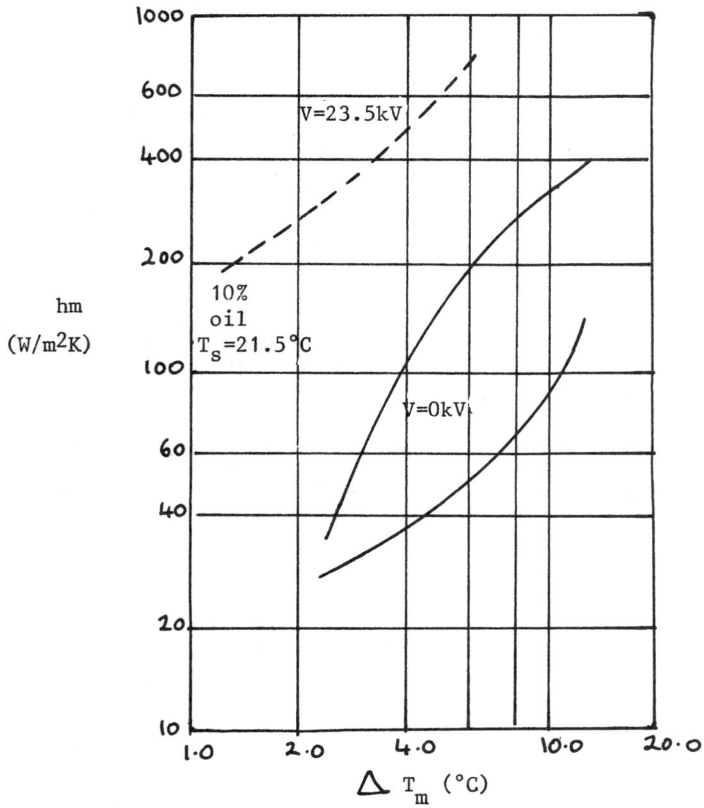

Fig. 4. P.H.G. Allen's Data on Electric Field Effects on Boiling of Refrigerant/Oil Mixtures.

intensification which can occur. The most recently reported work (7) concentrates on evaporation heat transfer with R114 as the working fluid, as part of a programme aimed at the development of full scale EHD enhanced evaporators and condensers.

The principal observations made during the recent programme were that EHD eliminated boiling hysteresis and enhanced nucleate boiling when used in systems with pure R114, and, most interestingly, these benefits were repeated when fields were applied to heat exchangers where a mixture of R114 and oil, a common situation, was used.

Fig. 4 shows the results obtained in the latter instance, where a quantity of oil (10% by weight) had been added to the system.The arithmetic mean heat transfer coefficient is plotted against the temperature difference between the wall and saturation. While more dramatic improvements were achieved with pure working fluids, the enhancement technique, although here involving potentials of up to 23.5kV, gave very good heat transfer coefficient improvements and also dramatically reduced foaming of the mixture.

While the voltage potentials appear high, the associated currents in this case were less than 10 micro-amps. In the case of a 'lo-fin' type of evaporator tube, the enhancement created by an expenditure of about 0.25 W is of the order of 200 W - a significant gain.

Others have found that enhancements need not be limited to liquids or two phase flow. Corona wind enhancement of convection in air has been investigated, and such a device has been produced for cooling of compact electronic arrays.

7. CONCLUSIONS.

In this paper a number of techniques for enhancing heat transfer have been discussed. These are not put forward as a solution to all the problems of low temperature heat recovery, some of which will be much better tackled by new materials, particularly plastics, and superior design tools. However, I believe that some of these enhancement techniques have an important role to play in making heat recovery equipment more cost-effective.

In a survey I carried out of US Patents published during one month of 1987 for a Journal I edit, of 30 patents relevant to heat recovery, no fewer than 10 were associated with plant miniaturization and /or process intensification. Examples included the use, by du Pont de Nemours, of low frequency, high amplitude vibrations in a plastic tube gas-liquid heat exchanger, a proposal by Sundstrand for a centrifugal heat exchanger incorporating liquid impingement jets, and a laminar flow shell and tube counterflow heat exchanger of novel design which appears to challenge the volumetric heat transfer coefficients of the plate unit discussed earlier.

I hope that this paper may stimulate further thoughts in these directions.

REFERENCES.

(1) FORSSMANN, W.G. et al. (1981). Normal and Pathologic Physiology of the Skin III In: Handbuch der Haut- und Geschlechtskrankheiten, Springer-Verlag, Berlin.
(2) COTTER, T.P. (1985). Principles and prospects for micro heat pipes. Los Alamos Laboratory Report.
(3) HRYNISZAK, W. (1958). Heat Exchangers. Applications to Gas Turbines. Butterworths, London.
(4) FINLAY, I.C. (1967). Heat transfer enhancement by addition of a liquid phase. Nature, Vol. 214, No. 5086, p. 430.
(5) CROSS, W.T. and RAMSHAW, C. (1986). Process intensification: Laminar flow heat transfer. Chem. Eng. Res. Des., Vol. 64, pp 293-301.
(6) US Patent No. 4553408. (1985). Centrifugal Heat Pump. Assignee, ICI plc.
(7) ALLEN, P.H.G. and COOPER, P. (1987). The potential of electrically enhanced evaporators. Proc. 3rd Int. Symp. on Large Scale Application of Heat Pumps, Oxford. BHRA, Cranfield.

ENERGY RECOVERY BY MECHANICAL RECOMPRESSION OF HYDROCARBON VAPOUR

J.P. LIVERNET
Sté RHONE-POULENC CHIMIE - Usine de CHALAMPE

This installation enables to save 34,6 T/h steam, pressure 6 bar, nominal production unit ; it is integrated into a cyclohexanol-one production unit, capacity 150,000 tpy, which has been operating since 1972 at the RHONE-POULENC CHIMIE plant in CHALAMPE (68 FRANCE).

1. RATIONALE OF THE PROJECT

The basic idea is to recover the calories contained in the cyclohexane vapour issued by the head of a distillation column. Until now, these calories were wasted : they were released during vapour condensation (passing into liquid state) and discharged into ambient air. Therefore, the project consisted in directly repressuring this vapour from the column head, in order to raise its temperature to such a level that it could be used as heating fluid in the column boilers.

This idea has already been widely applied with steam. The originality of this project lies in the use of a hard to handle processed fluid (flammable and corrosive due to traces of acid). The difficulty was to develop suitable mechanical equipment :

- high power : 4,200 KW
- absolute need of perfect tightness of the gas circuit
- problem of thermal expansion considering the thermal level and the selected metal (stainless steel).

With regard to the risks of corrosion, the whole equipment, including the compressor, is made of stainless steel.

1.1. Design studies

The design studies were performed by the Rhône-Poulenc Central Engineering Department, in cooperation with the plant Engineering and Operation Services. A detailed mock-up (1/33) was used for detail studies ; it was therefore possible

- to use optimally the available space
- to avoid pipework alterations at start up
- to train operating personnel.

Moreover, this mock-up was used by the building teams for the unit assembly.

1.2. Building and assembly work

It was performed from April 1983 to September 1984 within a production unit.

Assembly by welding being excluded (owing to the high inflammability of cyclohexane contained in the unit), four shut-downs of the unit were necessary to complete the installation erection.

Start up of the unit : January 85 after all the compressor tests.

1.3. Cost

The total cost was 41,000 KF in January 85. This project had received subsidies for the innovatory aspects from the commission of the European Communities and from the French Agency for Energy Control.

1.4. Energy savings

Many measurements have shown that this mechanical compression of organic vapours, for a 7000 hr/yr operation, brings about :

. for the plant : savings of 5,500 TOE/yr (TOE : equivalent metric ton of oil)

. for the domestic Energy : a substitution of hydrocarbon of 16,500 TOE/yr by 40,000 MWH/yr of electrical power including about 50 % for the compressor and 50 % to make up for the plan self-production.

The project payback mainly depends on the fuel oil and power prices and is expected by about 5 years without subsidies.

2. PROJECT DESCRIPTION

2.1. Sketch diagram description (see diagram enclosed in appendix)

2.1.1. Main circuit

The product to distillate is fed at mid height of column 22.02 ; in the column base, the concentrated heavy product mixture (cyclohexanol + cyclohexanone + heavy products) is drawn off by pumps and sent into the process sequence. The calories in the base are supplied by 3 boilers 26.12, 26.13 and 26.14.

The vapours issued by the column head (nearly pure cyclohexane + a little acid water) are superheated by about 12°C (exchanger 27.33), then compressed by compressor 60.12 before being sent to the three boilers (this superheater is made necessary by the low adiabatic compressibility coefficient of cyclohexane).

Most of the cyclohexane condenses in the boilers ; cyclohexane condensates and non condensed vapours are directed to tank 11.40, out of which comes :

a) from the top, about 7.5 % of the vapour entering the boilers ; this purging, which corresponds to the excess of calories in the system, ensures overall pressure regulation and allows fractional condensation of cyclohexane in the air condensers, thus draining off the little acid water contained in the vapours.

b) from the bottom, the cyclohexane condensates which, after going through the cyclohexane vapour superheater, are flashed in tank 11.38 to produce :
   - . recovery of the vapour which is recycled in the compressor
   - . liquid cyclohexane which goes to tank 10.01, where it is drawn off by pumps to ensure the 22.02 column reflux, the rest being recycled in the process.

2.1.2. Secondary circuits

. The system is started by the two boilers 26.01 and 26.02 with steam.

. At start up, during the heating of the compressor circuits, all the low points are purged through tank 11.37.

. The compressor is protected by an anti-pumping device.

2.1.3. Operation without heat pump

The possibility remains to operate without a heat pump : in that case, boilers 26.01 - 26.02 are used with a steam supply ; the hydrocarbon vapours issuing from the column are sent towards air condensers.

The mode of operation can be rapidly changed by using a set of automatic valves.

2.2. Main features of the equipment

The whole equipment is made of stainless steel. All metal fabricated devices are designed by Rhône-Poulenc.

Compressor 60.99

. Trade mark : ALSTHOM.ATLANTIQUE

. Type : single wheel compressor with multiplier and sliding lock paddles on suction
. compression ratio : 2.5
. Capacity : 80,000 m^3/h at 1.2 bar absolute pressure and t = 100°C
. Rotation speed : 3,500 rpm
. Maximum absolute power : 4,200 KW
. Shell diameter 2,5 m, wheel diameter : 1.25 m
. Oil tank integrated in the machine frame work.

Compressor view

Boilers 26.95/26.96/26.97 : tubular exchangers
. The 3 devices are identical
. Features of one device : surface : 1,097 m^2, length : 4 m, 3,450 tubes diameter 1".
. Shell diameter : 2 m.

2.3. Special features of the installation
. Compressor installed on a concrete socle resting on resilient slabs, in order to avoid wave propagation.
. The gas pressure is constantly maintained at a pressure higher than atmosphere, to avoid any air inlet.
. Fire protection by deluge network.
. Stationary Hydrocarbon detectors.
. Compressor insulation by automatic valves with remote control in the control room.

3. FUTURE PROSPECTS AND DEVELOPMENT

The devices of vapour mechanical recompression whose technical and financial interests have been pointed out, are well established in the food industry on the steam.

Chalampé experience has been achieved on vapours of hot and corrosive organics, using a high technology. This technique is full of promise, in particular to optimize the operation of refining and concentration columns which requires a lot of energy.

General View
Column - boilers - Pipe
and local compressor

Mock up (1/33)

COLONE 4 MECHANICAL RECOMPRESSION

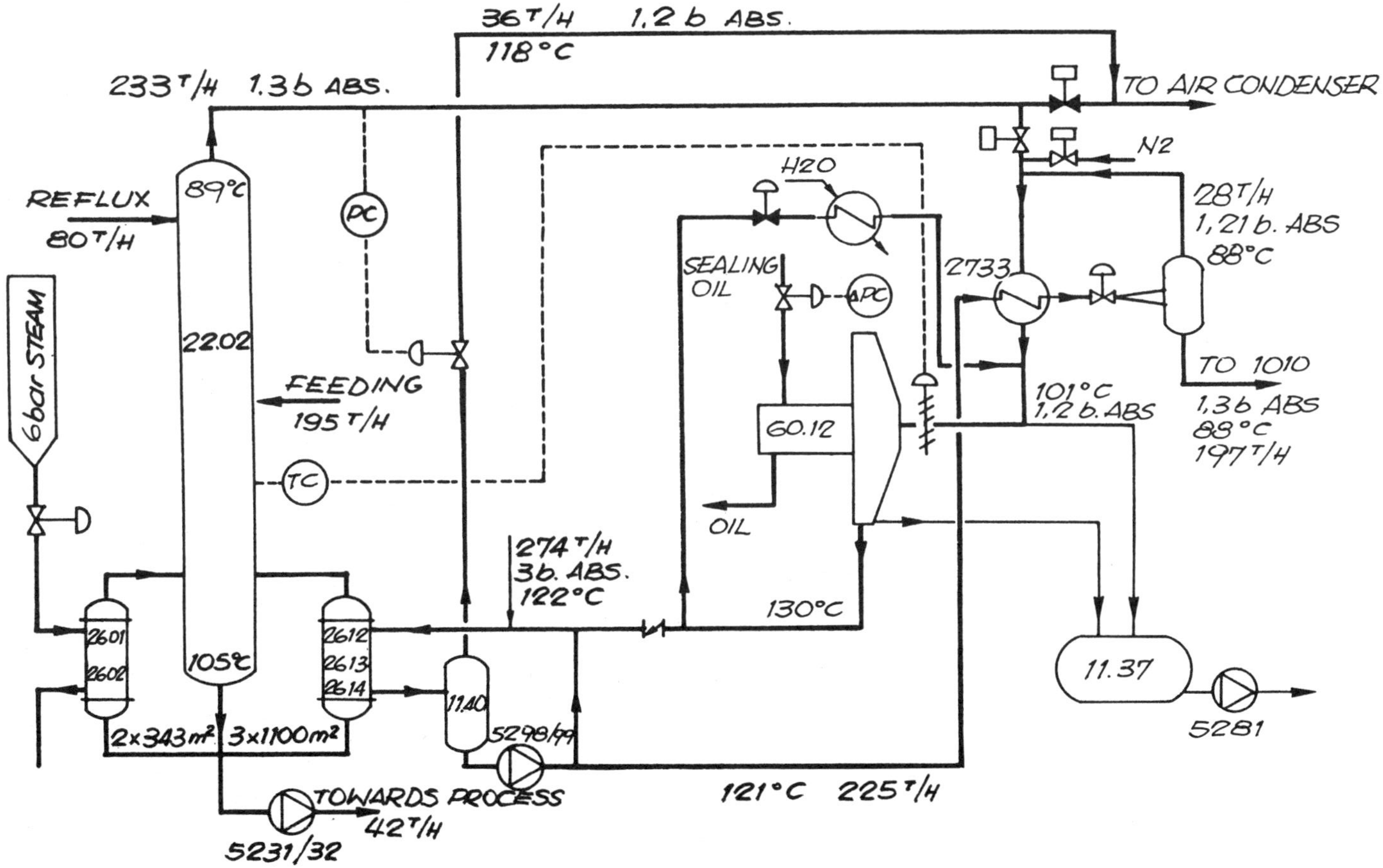

HEAT EXCHANGERS IN PLASTIC

J. HUYGHE
GRETh* General Manager

1. INTRODUCTION

A recent French study on the use of energy in industry showed that most of the energy was consumed at low temperatures, i.e. between about 50 °C and 200 °C.

The recuperation of thermal energy in this interval is of great interest. We are aware that the main difficulties encountered in this task are the fouling and the corrosion of the surfaces of the heat exchangers in industrial processes which, combined with the investment and the expenses involved in the upkeep of the heat exchangers, often prevents this recuperation.

Due to recent advances in the development of plastics and the knowledge acquired in the techniques of their use, plastics can now be used for the manufacture of heat exchangers. As a result of qualities inherent to these materials, solutions to the difficulties mentioned above can be found.

The way was opened in 1965 by the American firm DUPONT DE NEMOURS with the development of a polymer known commercially under the name TEFLON. DUPONT DE NEMOURS was the first to manufacture heat exchangers in plastic and is still one of the world leaders in this field. However, increased efforts in research and development have been undertaken for the past several years in Europe, and today European heat exchangers in plastic are appearing on the market.

Here, after going over the principal advantages and disadvantages of plastics, we will examine the different possibilities for their use in heat exchangers and describe some examples of recent achievements.

2. PROPERTIES OF PLASTICS

The principal characteristics (advantages and disadvantages) of plastics will be seen from the angle of their use in heat exchangers.

2.1. Weight

The specific weight generally ranges from 900 to 2200 kg/m^3, or 4 to 5 times less than most metals. Exchangers made of plastic will, therefore, be lighter, for the same volume, than those made of metal.

2.2. Surface aspects

Parts for the manufacture of heat exchangers, especially tubes or plates used for heat exchange surface, are particularly smooth. Friction factors and, therefore, pressure losses are lower than for metals at identical dimensions and flow rates.

Moreover, the wettability is very low. Therefore, water vapor condenses on the plastic walls in the form of droplets rather than a continuous film. This increases the heat exchange coefficients during condensation compared to metallic exchange surfaces.

Lastly, mineral and especially organic deposits adhere much more

---

* GRETh : Groupement pour la Recherche sur les Echangeurs Thermiques created by A.F.M.E. and C.E.A. to support industry of heat exchangers. Address : GRETh - CENG - 85 X - 38041 GRENOBLE CEDEX (FRANCE).

difficultly to the walls. This lessens the risk of fouling and makes it easier to clean the exchangers.

2.3. Chemical resistance

Most plastics are resistant to corrosive fluids such as organic or mineral acids, oxidizing agents, hydrocarbons, chlorine, bromine and their compounds.

The most remarkable plastics are PTFE** and PVDF**, fluorine chain thermoplastic polymers.

2.4. Mechanical resistance

The specific resistance of plastics, the ratio of the mechanical resistance to the density, is the highest of all materials.

However, the mechanical resistance, which decreases rapidly as the temperature rises, is about ten times less than that of metals at ordinary temperature : the resistance to traction varies from 10 to 100 MPa for basic thermoplastics and from 200 to 800 MPa for reinforced plastics. The elasticity modulus is about 3000 MPa (200,000 MPa for steels). This situates plastics between wood and rubber. Fiberglass on carbon loads improve greatly the mechanical properties.

Finally, the resistance to erosion and abrasion of plastics intended for use in exchangers is often superior to that of metals.

2.5. Thermal expansion

Thermoplastic polymers dilate about ten times more than metals. It is therefore necessary to allow adequate clearance in the metal-plastic assemblies.

2.6. Thermal conductivity

The thermal conductivity is from 100 to 300 times less than that of metals commonly used in heat exchangers : the thermal conductivity coefficients are generally between 0.1 and 0.4 W/m^2°K. This disadvantage is often cited by opponents to the use of plastics in heat exchangers, but we will see below how this disadvantage can be overcome.

HDPE** has the highest value for thermal conductivity : 0.4 W/m^2°K. By adding conducting loads (carbon fibers, powdered graphite or aluminium), the thermal conductivity coefficient can be doubled or tripled.

2.7. Resistance to humidity

Some plastics have a tendency to absorb water (1 to 4 % of their volume) which can provoke undesirable swelling.

2.8. Inflammability

All plastics are inflammable to different degrees.

2.9. Ageing

This is the least well-known phenomenon. The main factors in ageing are high temperature, mechanical stress, some chemicals agents and ultra violet rays. The lifetime of some linear chain thermoplastic polymers can be greatly increased by crosslinking either chemically or through irradiation. Fig. 1 shows the difference in behavior of HDPE** vs time. It can be seen that by limiting the temperature and the stress, this material has a lifetime of several decades.

** See nomenclature, page 90.

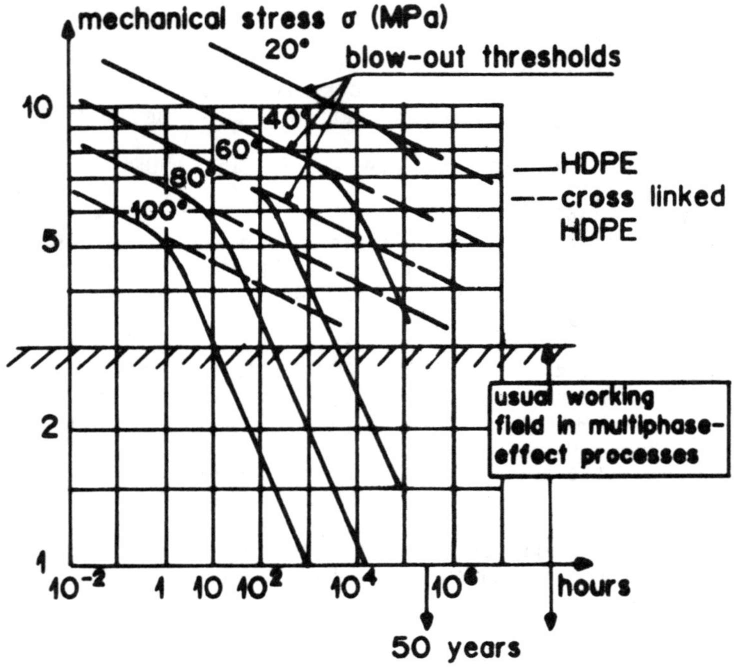

Fig. 1 : Test results on the lifetime of PEHD

2.10. Temperature

The mechanical properties degrade rapidly as the temperature rises. The maximal operating temperature depends on the mechanical stress and on the projected lifetime.

Manufacturers of basic materials generally indicate the maximal operating temperature. For example :

PTFE (Poly Tetra Fluor Ethylene)	250 °C
PS (Poly Sulfone)	160 °C
PVDF (Poly Fluoride Vinyldene)	140 °C
HDPE (High Density Poly Ethylene)	110 °C
PP (Poly Propylene)	80 °C
PVC (Poly Vinyl Chloride)	60 °C

2.11. Price

The price of plastics varies considerably depending on their nature. It ranges from 1 to 2 Ecus/kg for the least expensive (PP) to 25 Ecus/kg for PVDF to 70 Ecus/kg for PTFE. Nevertheless, the price of low temperature thermoplastics remains low compared to that of metals.

2.12. Transformation

Plastics can be easily transformed. Molding, extrusion, injection, thermoforming, soldering and glueing are commonly done today, facilitating their use in the manufacture of heat exchangers.

3. THE USE OF PLASTICS FOR HEAT EXCHANGE SURFACES

For the thermal engineer, the major disadvantage of plastic is its low thermal conductivity (cf. § 2.6.) even though this can be compensated by plastic's low thermal resistance to fouling or its better thermal exchange coefficient during condensation (cf. § 2.2.).

It is well-known that the determining parameter for the exchange surface of a heat exchanger is its overall heat transfer coefficient U defined by the ratio :

$$\frac{1}{U} = R_1 + R_2 + R_p + R_f \qquad (1)$$

where R_1 and R_2, the thermal resistances of fluids 1 and 2 respectively, are calculated by the classic laws of heat transfer (generally forced convection), and where R_p is the thermal resistance of the exchange wall due to conduction, defined by the ratio e/λ (in a plane wall) ; e is the thickness of the wall and λ the thermal conductivity of its material. R_f is the thermal resistance to fouling which will be neglected here for reasons of simplicity.

Equation (1) shows that in exchangers made of plastic, for which the thermal conductivity is much lower than for metals, the thermal resistance of wall R_p is much higher, except for the case where a plastic much thinner than metals is used. However, the respective values of the thermal resistances R_1, R_2 and R_p must be taken into account, hence the characteristics of the circulating fluids and the functions of the exchanger as well. It is well-known that heat transfer by convection in a gaseous fluid, for example, is generally weak. Therefore, the thermal resistance of the wall is of little importance in determining U.

Fig. 2 shows the ratio U/U_0 in function of U_0. U and U_0 are calculated by formula (1) with $R_p = e/\lambda$ for U and $R_p = 0$ for U_0. This figure illustrates the influence of R_p on U for two different materials : stainless steel 1 mm thick and three different thicknesses of HDPE : 1 mm, 0.3 mm and 0.05 mm.

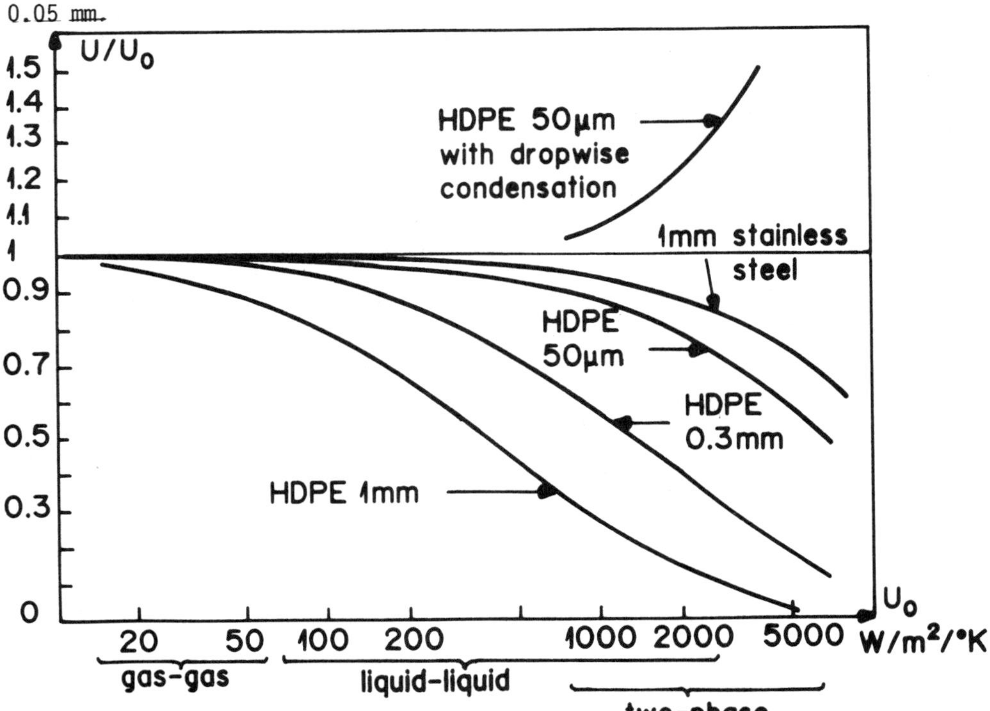

Fig. 2 : Influence of R_p on U

When HDPE is used for a gas-gas heat exchanger where the overall heat exchange coefficients are from about 20 to 50 $W/m^2°K$, there is little difference between an exchange surface in stainless steel and an exchange surface in HDPE 0.3 mm or even 1 mm thick (there is a 10 % discrepancy at 1 mm in that last case). However, for a liquid-liquid heat exchanger where the heat exchange coefficients are about 1000 $W/m^2°K$, the thickness of the HDPE must be minimized (0.3 mm is the maximum acceptable : a 30 % discrepancy). For use in a condenser or an evaporator where the overall heat exchange coefficient is about 5000 $W/m^2°K$, a value of the heat exchange surface comparable to that of a heat exchanger in stainless steel can be obtained only with a 50 µm thickness of HDPE. When thin HDPE is used for a water vapour condensation surface, the ratio U/U_0 can go beyond 1 due to the excellent heat transfer by condensation of HDPE and its poor wettability which brings about dropwise condensation (cf. § 2.2.). The corresponding curve in Fig. 2 is based on experimental results.

For this study, the heat exchangers will be classified into three types according to their use : gas-gas heat exchangers, liquid-liquid heat exchangers and two-phase flow heat exchangers (evaporators or condensers).

4. DIFFERENT TYPES OF HEAT EXCHANGERS IN PLASTIC

Heat exchangers in plastic can be made in three ways : shell-and-tube heat exchangers, plate heat exchangers and flexible sheath heat exchangers. They will be classified in the types of use defined at the end of § 3, depending on their respective characteristics.

4.1. Shell-and-tube heat exchangers

These heat exchangers are of classical type. The main elements are the tubes, the shell, the tube sheets and the intermediate baffles. They can include one or two passes. Some heat exchangers commercialized by European firms are monoblock, all plastic, and contain no joints. The polymers used are generally PP, PS or PVDF. The shell can be made of the same material as the tubes or of fiberglass resin.

The first heat exchangers of this type made in Europe used tubes with relatively large diameters, 10 to 25 mm, that were about 1 mm thick. Their thermal performance was therefore weak, corresponding to overall thermal exchange coefficients of about 60 to 150 $W/m^2°K$. They were bulky and also expensive when high quality polymers were used. They were, however, essential in some cases where their lifetime was longer than that of metallic heat exchangers.

Today, due to progress in the transformation of plastics (especially in extrusion processes for very thin tubes), a new type of heat exchanger using small diameter (1.5 to 4 mm) and thin (0.15 to 0.35 mm) "capillary" tubes has been developed in Europe (notably ENKA, VICARB...). Because of the tube dimensions, these heat exchangers perform better than those of the first generation (overall heat transfer coefficients of 500 to 800 $W/m^2°K$). Above all, they are more compact : 400 tubes can be housed in a tube-sheet, 90 mm in diameter (See photograph below).

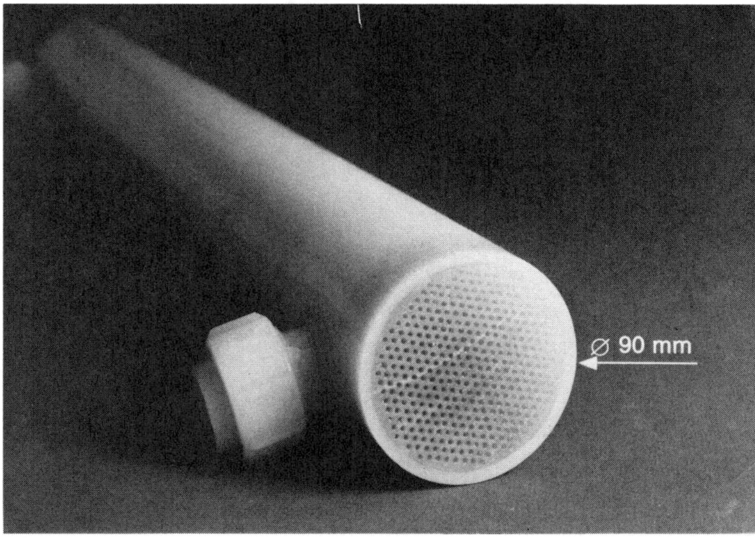

Photo 1 - Shell- and tube heat exchanger with "capillary" tubes

A very high heat transfer area per unit volume can be obtained : 200 to 600 m^2/m^3, or 10 to 20 times more than previous exchangers. Plastic heat exchangers are becoming, therefore, more economical compared to metallic heat exchangers (which are generally made of a "noble" metal like Tantale or Hastelloy) and offer the guarantee of a much longer lifetime. The size of plastic heat exchangers is, however, still limited : heat exchange surfaces of 30 to 50 m^2 maximum.

In addition, the use of small diameter tubes brings about a low-speed, laminar flow regime in the tubes, leading to low values for pressure losses.

The temperature performances depend on the material used : PP (up to 80 °C), PVDF (up to 140 °C) or PS (up to 160 °C). Depending on the temperature used, they are guaranteed from a few bars to 10 bars. The tubes, the tube-sheets and the shell are often made of the same material. The tube/tube sheet link can be made by glueing for use at low temperatures or soldering for use at higher temperatures. The use of this new type of apparatus, generally in a liquid-liquid heat exchanger, is becoming more and more frequent in the chemical, pharmaceutical, agroalimentary and electronics industries.

4.2. Plate heat exchangers

Plastic can be used for the exchange surface in plate heat exchangers (which are generally used in liquid-liquid heat transfer) if it is thin enough to conserve values for the heat transfer coefficients comparable to those of metallic exchangers (See Fig. 2). A plate heat exchanger using thin sheets of plastic for the exchange surface has been developed in France and is beginning to be marketed (1). See photograph below.

Photo 2 - Plastic-plate heat exchanger

This exchanger consists of a stack of plastic frames (about 5 mm thick) separated by thin sheets of plastic (about 0.1 mm thick). In certain models, thin sheets of stainless steel or titanium (0.1 mm thick) can also be used. The entire device is held, joint free, in a clamp stand by means of tightening bolts like a plate heat exchanger with ordinary joints. The seal is assured by adequate tightening of the bolts. Plastic grids are placed inside the frames ; they maintain the gap and create turbulence favorable to forced convection heat transfer across the heat exchange surface. At the ends of the frames, distributors assure the inlets and outlets for each fluid. The entire heat exchanger is made of plastic except for the clamp stand and the tightening bolts which remain in metal. This characteristic makes the heat exchanger highly resistant to corrosion and inexpensive.

It should be noted that many problems during assembly and operation are avoided by the total absence of joints. The apparatus is easy to disassemble for cleaning and can be reassembled as it is, due to the absence of joints and glueing.

This heat exchanger can be used for liquid-liquid applications or as a condenser. The maximal operating temperature is, at the present, 60 °C, and the nominal pressure is about 6 bars (2 bars between circuits, on each side of the plastic sheets). It is commercialized for exchange surfaces from 0.5 to 200 m^2 and flow rates up to 200 m^3/h. The material used is PP.

Its applications are principally in chemistry, aquaculture, geothermal applications...

4.3. Flexible sheath heat exchangers

These exchangers were initially intended to operate as condenso-evaporators. Their design results from the examination of the curves in Fig. 2 : in order to maintain acceptable performances of the heat exchangers operating in the two-phase flow zone, on the right of the figure, the plastic used must be very thin. In this way, a 50 μm thick exchange surface in HDPE (the most conductive of plastics yet known, see 2.6.) leads to a decrease of one tenth per cent in the overall heat transfer coefficient compared to a 1 mm thick exchange wall in stainless steel, all other things being equal.

A new type of heat exchanger has thus been developed from a bundle of tubes in very thin plastic (0.03 to 0.1 mm). These tubes are produced with a large diameter (20 to 40 mm) : from here on in, they will be called sheaths. These sheaths are pliable and are presented in a flattened form. They are furnished in rolls several hundred meters long. They are obtained by extrusion, and their price is modest.

The principle of a vertical falling film evaporator that uses these sheaths derives from the fact that the fluid under the highest pressure must circulate inside the sheaths to "inflate" them. As a result, as indicated in Fig. 3, the heating vapour goes through the sheaths while the solution to be evaporated flows in a film along the outside wall (the operation is the opposite of the falling film evaporators with metallic walls).

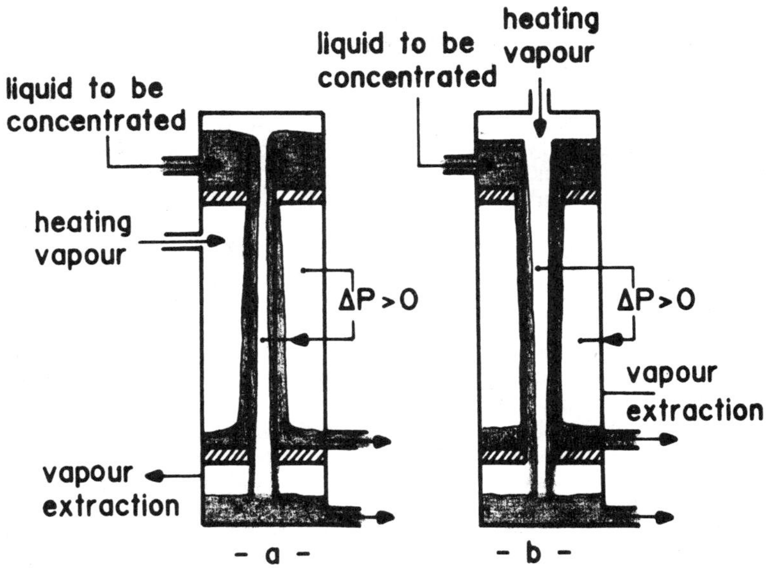

Fig. 3 : Comparison of the operating principles of a falling film evaporator with metallic tubes (a) and an evaporator with plastic sheaths (b).

This new design for a sheath evaporator heat exchanger is based on patented devices to fasten the sheaths to the tube-sheets and to distribute the solution to be evaporated. The parts are made of molded plastic. They are very low priced and installation is easy. The principle of that condenso-evaporator exchanger can be used in installations for concentration through multiple-effect evaporation (2) where the differences in pressure between effects and the operating temperatures are compatible with the stress values indicated in Fig. 1.

A multi-tubular two-effect pilot installation was set up by the CEA* at its salt water testing station in TOULON. The objective was the

---

* French Atomic Energy Commission

desalinization of sea water. Each evaporator was composed of 12 sheaths 3.5 m long and 30 mm in diameter. Different plastics were tested : HDPE, PVDF and polyamide. The pilot installation operated satisfactorily for four years. Temperature differences between effects ranged from 3 to 15 °C, and the temperature of heating vapour ranged from 100 °C to 40 °C. For sea water desalinization factories whose thermal exchange surfaces are in a "noble" metal (copper alloys or even titanium) due to problems of corrosion, it was calculated that a 20 % reduction in the cost of the installations could be obtained by using low cost HDPE for the exchange surface.

This process which passed its tests in sea water evaporation, can obviously be applied to any process of evaporation or concentration involving aggressive solutions. The risks of corrosion and therefore unplanned stops to production, can be reduced. In addition, because of the low price of the heat exchange surface, the number of effects can be increased, and performance can be improved. The KESTNER Company in France applied this concept to a process of evaporation by mechanical recompression of vapour (3). Due to the low cost of the plastic heat exchange surface, the size of the evaporator could be increased compared to that of a metallic exchange surface. The difference Δt between the temperature of the compressed heating vapour and the saturation temperature of the solution to be evaporated was diminished. As a result, the pressure rise ΔP to be provided by the compressor was considerably lowered, as were compressor expenses, directly proportional to ΔP. Thanks to this process called PLASTIREM, the KESTNER Company has announced a drop of 30 % in the investment price and 40 % in operating costs compared to classical processes. The COURTAULDS in Great Britain has also conducted research (aided by the C.E.C.) on a recompression process that uses an evaporator with very thin plastic tubes (0.13 to 0.2 mm) in a horizontal bundle over which the liquid to be concentrated is sprayed.

The design for a vertical evaporator has been simplified for use in gas-gas flexible sheath heat exchanger. Although the use of very thin plastic is not imperative for gas-gas heat exchangers (left side of Fig. 2), the advantages of such an exchanger are the resistance of the exchange surface to corrosion, its light weight, and its low price. It is in direct competition with glass tube exchangers, over which it has the advantage of its light weight and its sturdiness. But it also has the disadvantage of its temperature limits (about 140 °C for an exchange surface in PVDF). It can be used as a heat exchanger, recuperator in air conditioning installations in factories, for example, when the extracted hot air is loaded with corrosive vapour or in agricultural drying (cereal or vegetable dryers).

In this gas-gas sheath heat exchanger, as in the evaporator, the gas under the highest pressure must go through the sheaths to inflate them, and the gas under low pressure circulates outside the sheaths as indicated in the drawing in Fig. 4. The difference in pressure between the inside and outside of the sheaths must remain low. The limit depends on the operating temperature but is superior to the pressure of the gas circulating ventilators. The heat exchanger can be counter-current or cross-current.

1 : Removable steel flooring
2 : PDVF sheaths
3 : Sheath tension by springs
4 : Grating or vane

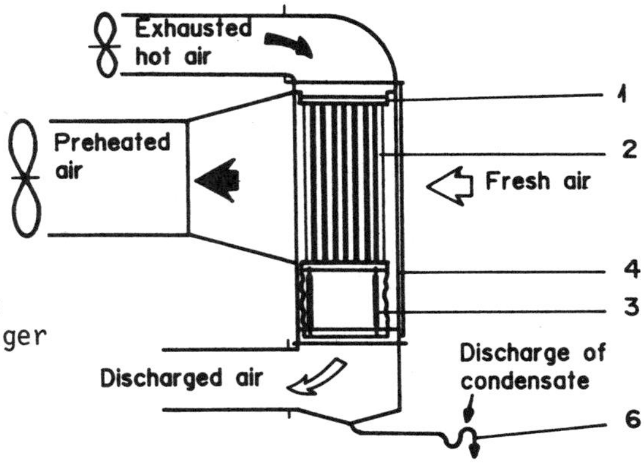

Fig. 4 : Plastic gas-gas flexible sheath heat exchanger Cross-current version

The French firm NEU has already set up in France and other European countries more than a dozen agricultural or industrial installations with this type of heat exchanger for gas flow rates varying from 10,000 to 100,000 Nm^3/h.

5. CONCLUSION

Some models of heat exchangers are available today for the recovery of thermal energy at low temperatures or for the improvement of industrial processes. From gas-gas heat exchange to liquid-liquid heat exchange, from evaporation to condensation, they fit most of our needs.

In addition to the well-known advantage of decreasing risks of corrosion, they often bring light weight and sometimes compactness (both of which are beneficial for links and support structures), a lower level of fouling, and a very competitive price compared to metallic heat exchangers. They have the disadvantage of being limited in resistance to temperature (except for expensive plastics).

In spite of the reticence of many potential clients, their use is spreading rapidly in industry. A 1985 study showed that the growth rate for the use of heat exchangers in plastic would be 30 % per year in France for the period 1985-90.

Moreover, it is certain that the rapid progress underway today in the elaboration of new, low-cost, basic materials that can be used at high temperatures (250 °C will probably soon be reached) combined with the improvement of the mechanical and thermal properties of plastics and a better knowledge of transformation techniques will accelerate this tendency.

REFERENCES

(1) VIDIL R., Les échangeurs à plaques et joints. Editions Lavoisier. Paris.
(2) LAURO F., HUYGHE J., Utilisation des matières plastiques comme surface d'échange de chaleur pour économiser l'énergie dans les procédés industriels de concentration par évaporation. Revue Physique Appl 17 (1982) 617-623.
(3) LELEU R., Compression mécanique de vapeur sur un évaporateur à surface d'échange à gaines plastiques minces. Colloque AFME "Maîtrise de l'énergie et recherche - Bilan et perspective". Paris, La Villette 3-10 décembre 1985.

VAPOUR COMPRESSION IN A BREWERY

E.Nolting, MAN Technologie GmbH

1. Introduction

Let me first present vapour compression in a broad framework. In the early 70s it was painfully brought home, especially by the Club of Rome, that ecology places insurmountable limits on technological society. Maintaining the balance in the biosphere is an essential condition for man's survival.

There are two chief factors that tend to upset this balance. The first is the direct cosumption of ecological resources, including the consumption of energy and air. The second disturbing factor is the continuing destruction of our environment by pollution. This includes air emissions of NOx, carbon dioxide, halogens, etc, as well as heavy metals in the soil.

A vapour compressor is a device which in connection with energy production helps to mitigate both factors. By making use of the heat in waste gases both the first problem - consumption of natural resources - and the second - environmental pollution - are countered, since fewer emissions are generated thanks to the more efficient use of energy. Nevertheless, every effort should be made to produce the energy for driving the vapour compressor with minimum emissions.

2. Description of the vapour compression plant

In modernizing their brewhouse the private brewery Dortmunder Kronen placed emphasis from the very outset on the environmental situation in a residential area and on saving energy. A gas-engine vapour compressor with an external boiler was therefore chosen as an economic system with extremely high energy saving and at the same time minimum emissions (Fig. 1).

For vapour compression a process-gas screw compressor, MAN model GHH SKÜL 321 driven by an MAN E 2542 E natural-gas engine is used. The second main component is the vapour Thermostar, which transfers the heat to the wort. This was supplied by the A. Ziemann Company, Ludwigsburg, which had also installed the two complete brewing lines with external boilers at the Kronenbrauerei nine years earlier. Integration of the vapour compression system in the existing brewhouse layout was undertaken in cooperation with the Ziemann Company. The gas-engine screw compressor was set up on a platform above the two coppers and was provided with a sound-absorbing hood. The noise level for the nearest neighbour was thus reduced to 40 dB(A). The complete assembly rests on vibration-absorbing mounts, so that no solid-body transmission is noticeable in the building. The vapour Thermostar with a diameter of 1.8 m and a height of 7 m was accommodated behind the two coppers. All of the plant components, particularly the pipes that come into contact with the wort, are of stainless steel. The gas engine is provided with a complete heat recovery system (Fig. 2), in which the engine heat - meaning the heat from the cooling water, oil and exhaust gas - is utilized. Furthermore, besides the usual heat utilization with inlet and outlet temperatures of 90 and 70 °C,

Fig. 1

Vapour compression set with waste heat recovery

Plant data:

Vapour volume	5650 kg/h
Vapour pressure	1/1.6 bar
Vapour temperature	100/114 °C
Engine output	166 kW
Engine waste heat	288 kW
Engine type	E 2542 E
Compressor type	SKÜL 321

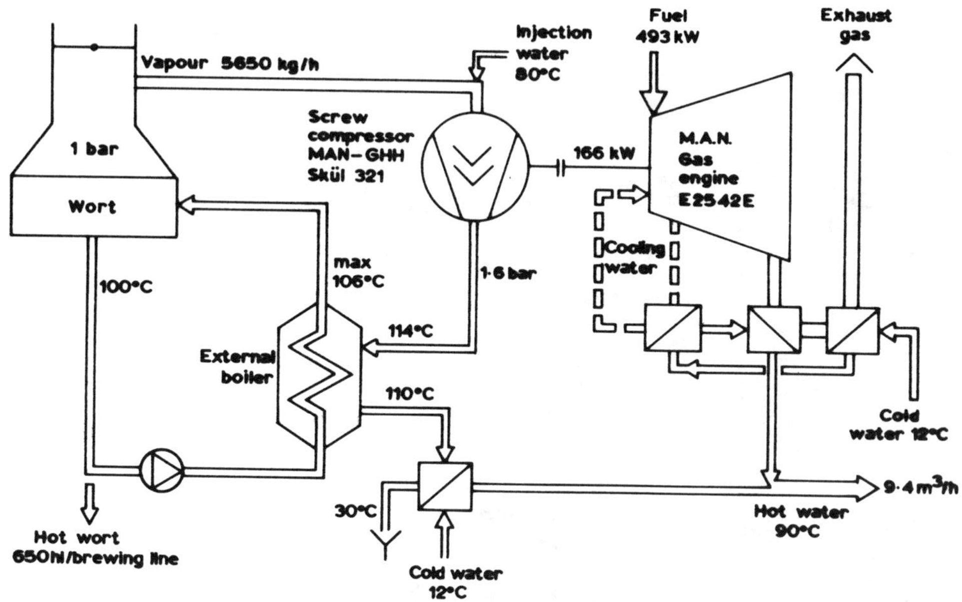

Fig. 2
Schematic diagram of vapour condensation with screw compressor and internal combustion engine
- Dortmunder Kronenbrauerei -

the low-temperature waste heat from the exhaust gas and radiated heat are recovered, as it can be used to heat the utility water of 15 °C. For this purpose a heat exchanger is located in the acoustic hood. The exhaust-gas heat is utilizied up to an exhaust temperature of about 60 °C.

Control and monitoring of the system was carefully coordinated with the brewers so that the quality of the beer is in no way affected. The control system is located on site and permits both manual and automatic operation. All of the essential data are transmitted to the control centre where they are incorporated in the brewhouse automation.

3. Function of the vapour compression system

Before the vapour compressor was installed the vapours generated during the boiling process were either released into the atmosphere or some was condensed and converted to hot water. Since the Dortmunder Kronenbrauerei has two brewing lines, alternative operation of the vapour compressor for both lines offered itself as a solution (Fig. 3). As soon as the wort in the copper has reached a temperature of 100 °C and vapours begin to form, the vapour compressor switches on, sucks off the vapour at a temperature of about 100 °C and compresses it to a temperature of between 110 and 120 °C at a pressure of 1.3 to 1.6 bar. This compressed vapour is then fed to an external boiler, the vapour Thermostar. In condensing, it heats the wort. It is thus part of the energy input for the boiling process, so that the losses that do occur are only those resulting from radiation through the thermal insulation of the overall system. 5,650 kg or, in the parlance of the brewer, 56 hl must be evaporated per hour. The vapour contains water droplets and impurities in the form of hop resins and oils. It was therefore advantageous to select a compressor design which could withstand exposure to these substances without impairment. The screw compressor with its moderate peripheral speed of about 100 m per second is ideal for such applications.

Because of its impurities, the condensate cannot be fed directly back to the process. However, to avoid losing the available heat, a condensate cooler was installed which, like the waste heat exchangers of the gas engine, is supplied with the cold utility water of the brewery. This contributes to an overall 92% efficiency of the natural gas used (Fig. 4). The saving in primary energy as compared to the former system is about 85% (Fig. 5). Since the brewing operation varies somewhat due to fluctuations in the starting material (e.g. seasonal variations in the barley sort) and different qualities of beer are brewed, it was essential for the vapour compressor in Dortmund to adapt to these changes. The main criterion was to adapt the load within a very broad range. With the gas engine this is possible in a continuous range from 50% to 100%. Here the system benefits from another advantage of the screw compressor, which has a constant pressure ratio over the entire volume range of 50% to 100%. Thanks to the design, there is no danger of pumping, as may occur with turbo compressors.

In particular, during the transition from wort heating to wort boiling, at which time vapours begin to form, a continuous smooth run-up of the vapour compressor over a period of 5 to 10

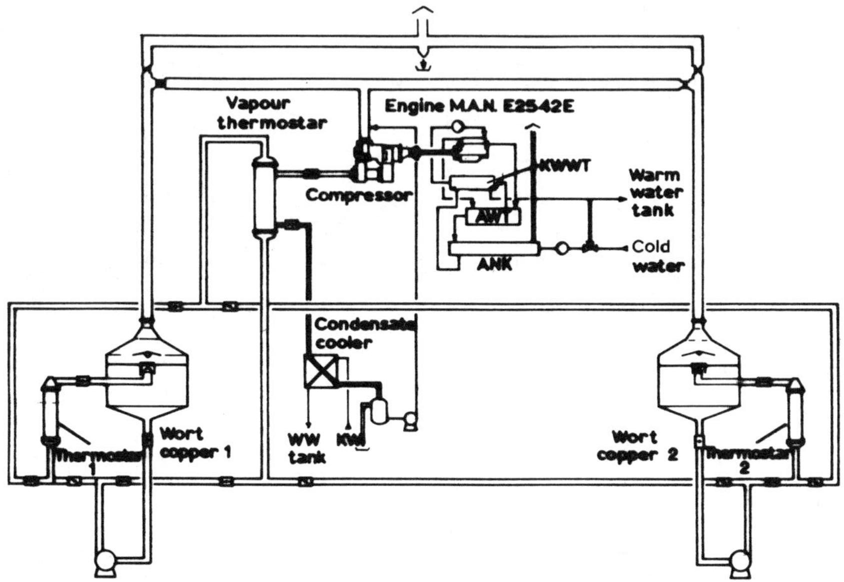

KWWT — Cold water heat exchanger   AWT — Exhaust heat exchanger
ANK — Condensate   WW tank — Warm water tank   KW — Cold water

Fig. 3 Heat recovery with vapour condensation.

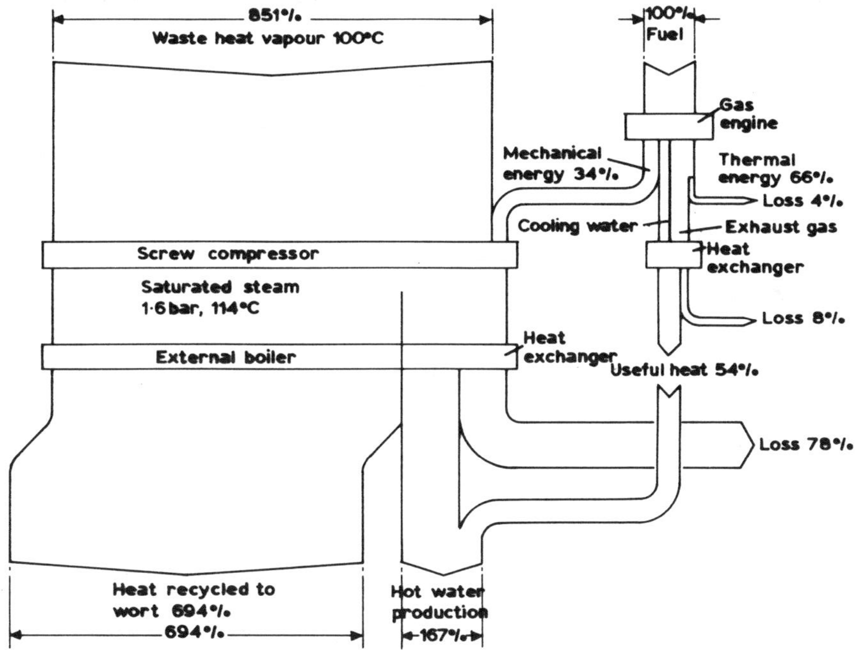

Fig. 4
Energy flow chart vapour condensation with screw compressor and internal combustion engine
- Dortmunder Kronenbrauerei -

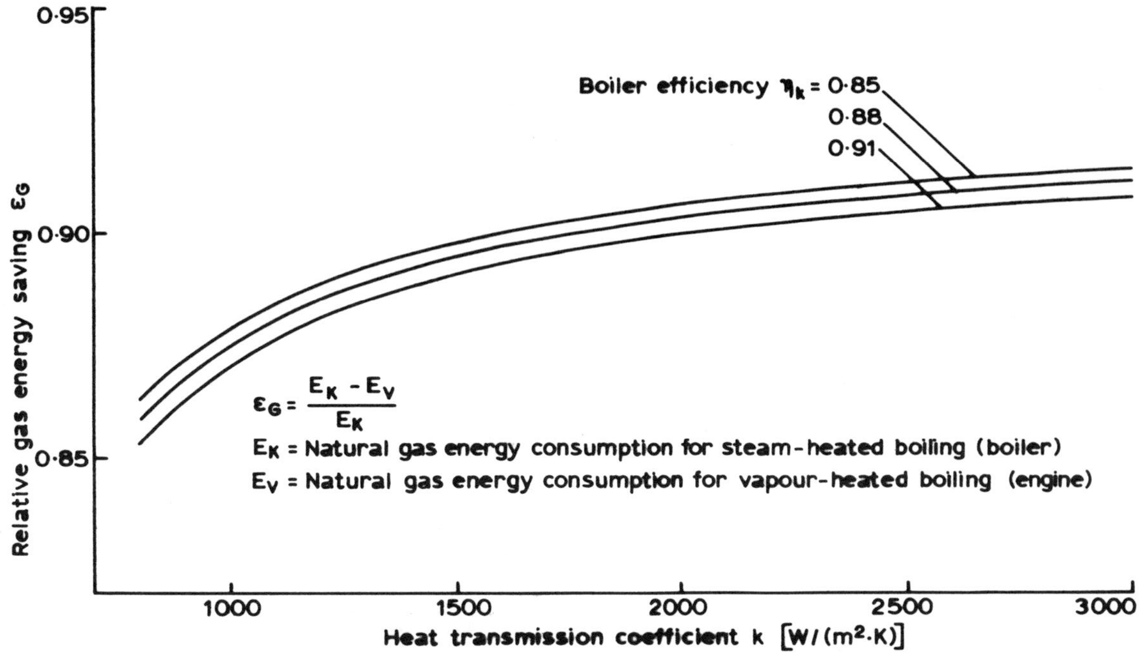

Fig. 5
Heat transmission coefficient of the external boiler k (W/m2 x k)
Data from Dortmunder Kronenbrauerei

Ek = Natural gas energy consumption for steam-heated boiling (Boiler)

Ev = Natural gas energy consumption for vapour-heated boiling (engine) minutes is advantageous. This can be elegantly solved with the speed control of the gas engine. Thanks to the closed circuit, the vapours no longer have to be released into the atmosphere, and atmospheric emissions are almost completely avoided. Certain residual emissions during run-up and run-down of the plant are unavoidable, but these remain below 5%. The organic compounds in the steam break down at temperatures above 140°C to form aromatics with a strong odor. This is prevented with the use of a screw compressor, since condensate is injected up stream the compressor and the temperature is maintained in the range of 120 to 130 °C.

4. Operating experience

The plant has meanwhile been in operation for 5,000 hours and has completely fulfilled its functions. All of the parameters concerning compression of the vapour volume, waste-heat utilization and fuel consumption correspond to the expected values. The reliability and serviceability of the gas engine have been confirmed. In short, the energy advantages of an internal combustion engine as compared to an electric drive have been fully realized and are not offset by downtime of excessive maintenance cost. In the initial phase the plant was mostly operated manually, so that the brewery staff could acquaint themselves with the process of switching from the boiler to the vapour compressor when the wort temperature of 100 °C was reached. Coordination of the control characteristics of the various servo valves was an essential point for stabilizing the operation. Thanks to the unstinting commitment of the staff at the Dortmunder Kronenbrauerei there has been no production downtime. Meanwhile the use of the vapour compressor has become part of everyday routine.

The performance of the system was measured at 12 points and recorded for one year (Fig. 6). The fluctuations in the operating parameters are quite small. It should especially be noted that the vapour Thermostar has to be cleaned only once a week, that is, after about 35 brewing cycles. The maximum rise in compression pressure due to impurities is 0.2 bar.

5. Conclusion

The goal was to realize a brewhouse modernization that substantially improves both emissions and the saving of energy. The solution at the Dortmunder Kronenbrauerei was a gas-engine-driven screw compressor. The results are a 95% reduction in vapour emissions, an approximately 85% saving in energy or natural gas, a 50% reduction in exhaust emissions and an amortization period of about two years.

Eight other plants in Europe - 6 in Germany alone - testify to the fact that this is not an isolated case. In principle, the operating results can be applied to all boiling processes in many branches of industry. In each case the influencing factors must be analyzed, the economic advantage examined and the best compressor system selected. The renewed rise in energy prices is sure to further stimulate interest in vapour compression.

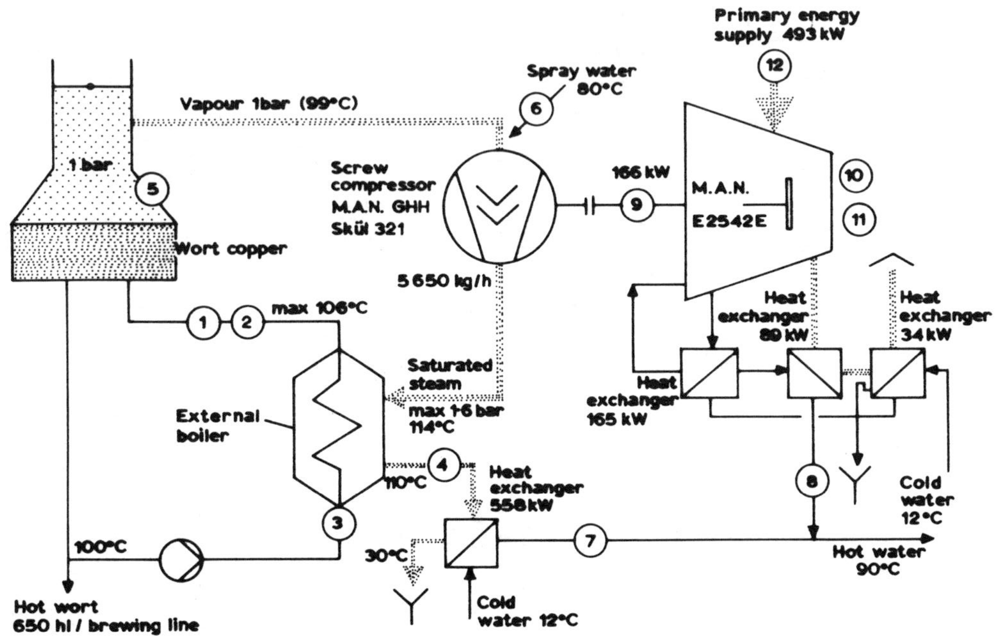

Fig. 6 Measurement diagram - vapour condensation.
Privatbrauerei Dortmunder Kronen

VALORIZATION OF RESIDUAL STEAM IN BRINE EVAPORATION

P.F. Bunge
Akzo Zout Chemie Nederland bv Research & Technology

1. INTRODUCTION

This paper describes an energy-saving-project in a salt evaporation plant, in which waste heat of very low temperature is being used.

In the case described here, the residual steam of a salt plant to be utilized has a temperature of 43°C, which is rather low.

By the installation of a specially designed brine evaporator it has become possible to use 40 % of this residual steam, which resulted in 5 % fuel cost saving for the salt plant of Akzo in Hengelo.

2. DESCRIPTION OF THE SALT PLANT

Salt is produced from brine by evaporation either in a multiple effect installation or in one effect with vapour recompression, or a combination of the two.

The two salt plants in Hengelo have both been designed with four evaporators, these are called "effects". See figure 1, page 2.

Brine, which is produced by underground dissolution of rocksalt followed by a chemical purification treatment is pumped through a series of preheaters and supplied to each of the evaporators. Evaporated brine, carrying the crystallized salt, is purged from effect to effect. In each effect, salt slurry is circulated over a heat exchanger. In the first effect brine is heated by condensing steam of approximately 130°C. This causes the brine in the evaporator to boil at a temperature of about 110°C. The temperature difference is made-up of the driving force needed in the heat exchanger, the boiling point elevation of the brine. There is also some temperature- and pressure-loss in the system.

The vapour generated in the first effect is used to heat the second effect. This must, of course, be operated at a lower temperature than the first one, which is realized by operating it at lowered pressure. This process is repeated from effect to effect.

The pressure of the vapour leaving the last effect is less than 0.1 bar and its condensing temperature is 43°C. It is condensed by direct cooling by water.

3. SAVING OF ENERGY

Most salt plants were built in a time when energy was still very cheap. Depending on factors like the price of steam, electricity, the possibility of steam-power co-generation, plant capacity and the price of equipment, the choice of the process, multiple effect or vapour recompression, was made and the heater surface and the number of effects were optimized. Akzo's salt factories in The Netherlands are all quadruple effect installations, with capacities ranging from 600,000 to 1.2 million tonnes per year.

If energy prices rise, it can be economical to extend a multiple effect installation with an extra effect upstream of the first evaporator. However, when the salt plant is linked to a steam-power co-gen-

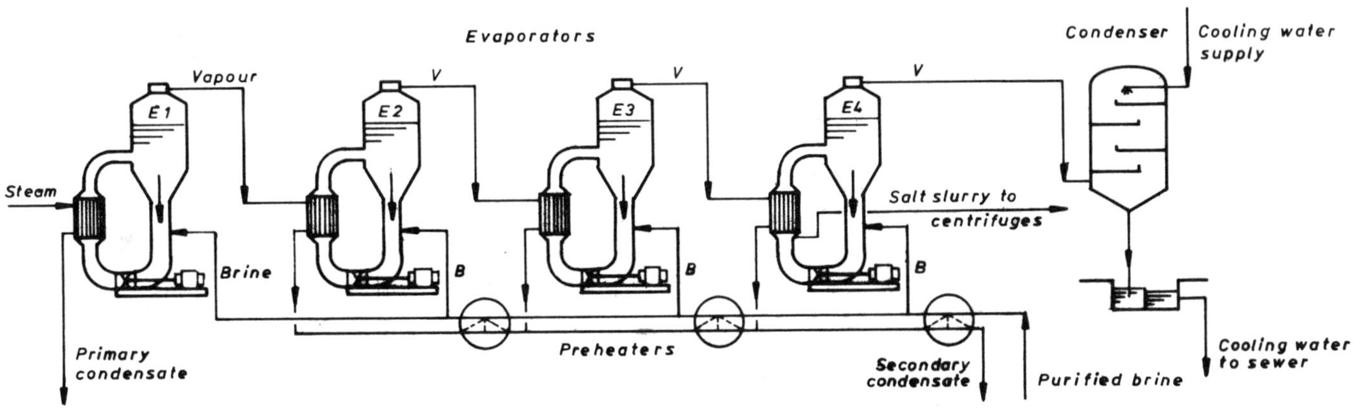

Figure 1. Salt production by multiple effect evaporation.

eration facility, the advantage of lower steam consumption is largely eliminated, since the additional evaporator will require steam at a higher pressure. Consequently less power will be generated. Moreover, in view of corrosion resistance, the evaporator would have to be constructed from copper-nickel alloys, which are very expensive.

As mentioned before, vapour from the last effect has a temperature of only 43°C. However, because of the large amount available, it contains a considerable quantity of heat.

The aim of the energy-saving-project was to design, construct, and operate on an industrial scale, a specially designed evaporator for the valorization of this vapour.

4. DESIGN CONSIDERATIONS

Before going into detail it might be informative to give an idea of the physical dimensions.

Figure 2. Evaporators in salt plant.

The last effect evaporator of this salt plant has a diameter of 7 m, and its vapour outlet at the top is 20 m above the floor. The vapour line has a diameter of 3 m.

For the design of the new evaporator there were several limitations:

- The temperature difference, available for heat transfer, boiling point elevation and temperature rise of the cooling water was only 18°C.
- investment should be relatively modest.

These limitations excluded the application of the generally accepted type of evaporation, i.e., a vertical evaporator with forced circulation heat exchanger and made from the usual materials such as copper-nickel and copper-nickel-cladding.

5. THE DESIGN

The temperature difference between residual steam from the fourth effect and cooling water is 18°C. About 6°C is available for heat transfer in the heater of the evaporator. The other 12°C is consumed by boiling point elevation, temperature rise of the cooling water in the condenser, and small losses in the system. However, considering how small the available temperature difference is, every tenth of a degree is important.

Conventional evaporators in salt plants are equipped with either an external or an internal heater, with forced circulation by means of a large pump, see figure 3.

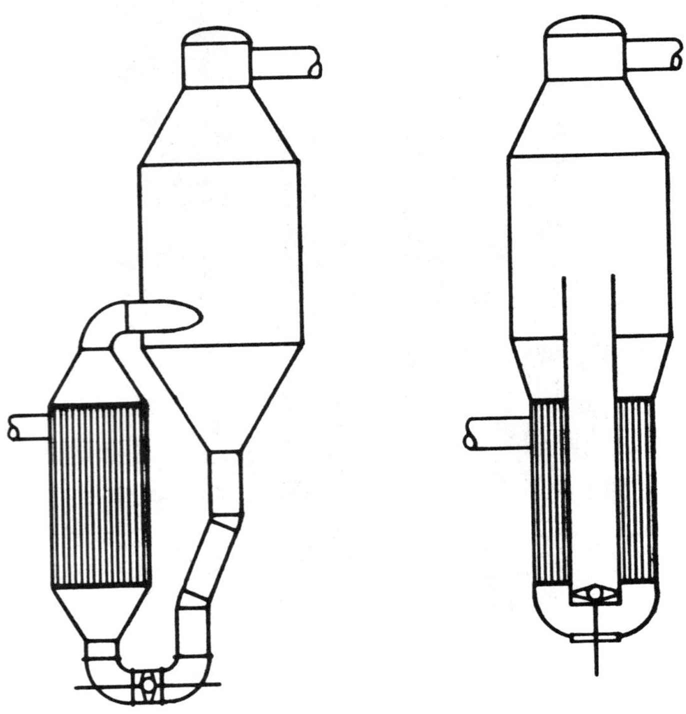

Figure 3. Evaporators with external and internal heat exchanger.

From our own observations and from experiments carried out at the University of Technology in Delft, it was known that an evaporator with an external heater shows a temperature loss in the order of 3°C by short circuiting of part of the heated fluid, back to the circulatory pump. In other words, this type of evaporator has limited "flash efficiency". Evaporators with an internal heater have much better flash efficiency.

The heating surface, needed for the new evaporator, was 5000 m², which is twice as large as the largest heat exchanger in the Hengelo salt plant. The construction cost of a conventional evaporator with such a large heater would be prohibitive.

The problems associated with the large heating area and high flash efficiency were solved simultaneously in a new evaporator design, see figure 4, page 6.

It consists of a horizontal evaporator body, equipped with two heaters of 2500 m² each, with a circulatory pump between them. Al though the flow pattern is much different from that of known designs, it was expected that the flash efficiency would be more or less equal to that of an evaporator with an internal heater.

Another point was that pressure-loss at the vapour inlet, which of course translates into temperature loss, had to be minimized. The Dutch Organization of Applied Scientific Research, TNO, were asked to carry out experiments. These experiments resulted in a hydraulic design of the vapour inlet with very low pressure-loss. TNO also carried out experiments to minimize the pressure-loss of the fluid in the pump elbow and heater inlet, resulting in a lower power consumption of the pump.

In order to keep the investment cost to an accetable level, it was decided to use carbon steel for the evaporator body as well as for the heaters, instead of copper-nickel. Taking into account the low operating temperature, the possibility of some corrosion was considered an acceptable risk.

Since it was not known whether this new type of evaporator could be operated as a crystallizer, it was decided to feed all the brine to the two salt plants in Hengelo to this evaporator, see figure 5, page 7.

The brine fed into the evaporator is not completely saturated. Although some solid salt is formed during evaporation, the crystal content of the large brine flow will remain very low. This is why this evaporator is called a "brine concentrator"

For carrying out experiments with a higher crystal content, it is possible to feed the brine to only one of the plants to the new evaporator, while maintaining the heat load.

6. PROJECTED COSTS AND SAVINGS

At the beginning of the project, the investment cost was calculated at Dfl. 8.5 million. Savings were calculated at 4430 tons of oil equivalent. At the time the project was proposed, the oil price was about Dfl. 500 per ton, and was expected to keep rising. So the annual saving was expected to be in excess of Dfl. 2.2 million.

7. INNOVATIVE ASPECTS

- Unlike known techniques, a horizontal evaporator body was chosen, so that a very large heating surface could be economically applied in a forced circulation circuit. To our knowledge there is no experience

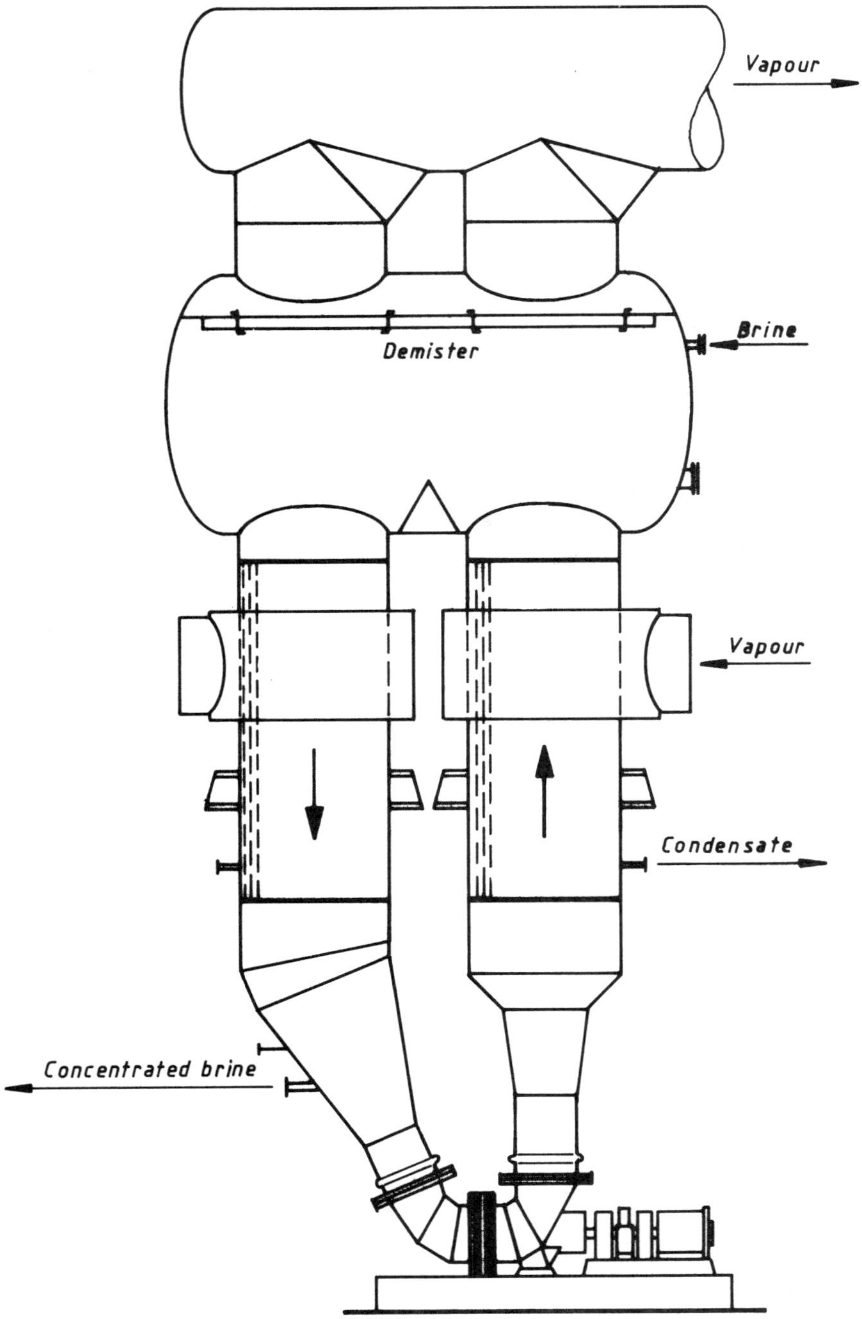

Figure 4. New evaporator design.

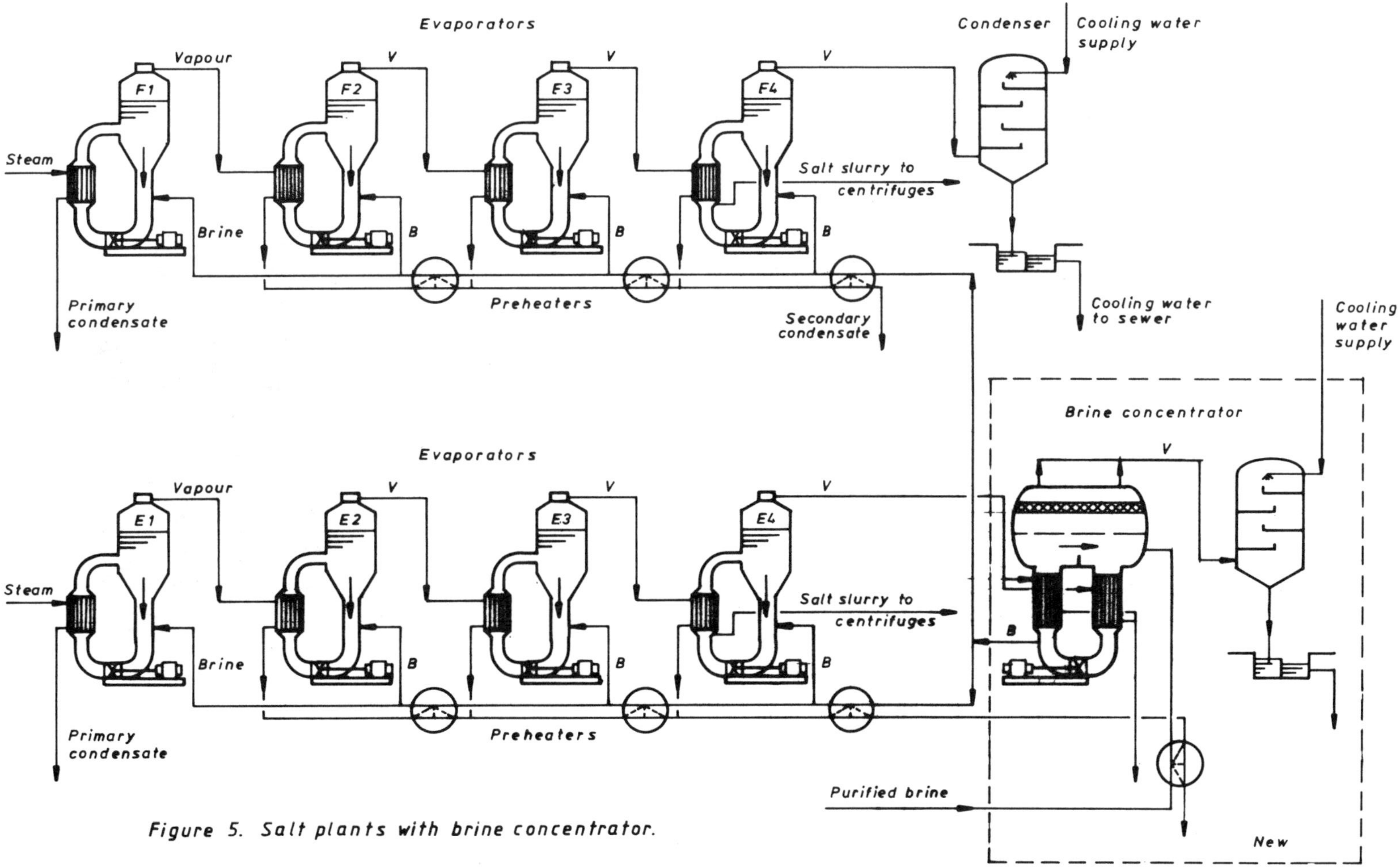

Figure 5. Salt plants with brine concentrator.

with respect to flash efficiency in a similar evaporator.
- Pressure-loss in the heat exchanger has been made extremely low by paying special attention to the construction of the vapour inlet, based on experiments in a model.

8. TECHNICAL AND ECONOMIC RISKS, TECHNICAL PERFORMANCE

Of course the technical risks have nothing to do with safety and environment. All technical risks have only an economic impact.

The risks involved with constructing this new type of evaporator, as considered at the moment when the project was being proposed, were:
- Flash efficiency may be lower than expected, so that either steam pressure has to be raised or the rate to which waste heat can be used will be reduced.
- Measures to reduce pressure-loss and to improve fluid distribution on the heater tubes will fail. See above.
- Slight corrosion on the vapour side of the steel heat exchangers, due to traces of salt from the fourth effect may reduce the heat transfer coefficient. See above.
- Salt crystals, which can be formed incidentally, e.g. during load alternations, could accumulate in the evaporator body, or lead to scaling, causing the need to flush the apparatus regularly. During flushing the apparatus must be by-passed.

The economic risk associated with these technical risks could lower the savings from the project by 20 to 25 %.

Figure 6. The brine concentrator.

In 1984 the brine concentrator was put into operation. Experience has shown that the flash efficiency is as expected, as is the case with the low pressure-drop. Corrosion does not occur. Incrustation of salt at the inlet of the tubes however, does occur. This problem was solved by applying short duration water injection at certain time intervals.

9. FINANCIAL RESULT

The calculated investment cost of Dfl. 8.4 million contained a 10 % allowance for contingencies. The project has been carried out for Dfl. 7.7 million. In contrast with the lower investment cost, savings exceeded expectations by approximately 10 %. In calculating these savings one has to take into account that, due to lower steam consumption, less electric power is generated, in this case about 2.5 MW, which should be compensated for at a public utility. At a public utility, electric power is produced at a lower efficiency than at the Akzo heat-power co-generation facility.

Compared to 1982, circumstances have changed. In the first two years of operation of the brine concentrator, the oil price was higher than the Dfl. 500.= per tonne in 1982, thereafter substantially lower. On the other hand, salt production was substantially higher than expected in 1982. The net result is that the profitability of the project is higher than expected. In terms of pay-out, the figure is 2.7 years after start-up, not taking into account the subsidy from the Commission of the European Communities.

10. OTHER BENEFITS

At normal conditions, no crystallization of any significance occurs in the brine concentrator. By by-passing part of the brine, while maintaining the heat load, the brine concentrator could be tested under crystallizing conditions. It was found that no clogging or accumulation of salt crystals took place. So as a side benefit of this project, know-how has been obtained as to the operation of a horizontal evaporator/crystallizer. The horizontal evaporator made it possible to install a very large heating surface with two heat exchangers in series, with a circulatory pump between them. The obtained know-how pertains to flash efficiency and crystallization. Plans have been worked out for the erection of crystallizing brine concentrators in other salt plants, but low energy prices at this moment have postponed these plans.

In principle the apparatus can be applied to any liquid evaporating duty. Apart from the production of salt one could think of the evaporation of caustic soda, of diaphragm cell liquor and of clarified juice in the production of sugar.

OVERVIEW OF THE EUROPEAN COMMUNITY RESEARCH AND DEVELOPMENT ACTIONS ON LOW TEMPERATURE HEAT RECOVERY

P.A. PILAVACHI

Directorate-General Science, Research and Development
Commission of the European Communities, Brussels

1. INTRODUCTION

The European Community has been involved in research on a European level in the field of heat recovery in the framework of three Energy Conservation Programmes.

The First Energy R&D Programme, launched in 1975, with a planned duration of 4 years, was followed by the Second Energy R&D Programme, announced during 1979, and also scheduled for 4 years. A Third Programme is currently under way, for completion in 1988. Projects within these programmes are supported under cost sharing contracts with the Commission. The total expenditure by the Commission on energy conservation projects during the First Programme was in excess of 11 MECU, 117 projects being supported. Funding for the subsequent Programmes was substantially higher, each being allocated approximately 27 MECU. However the number of projects supported in these Programmes, at 160 and 100 for the Second and Third Programmes respectively, reflects the trend towards increased project size and an encouraging growth in collaborative projects involving several Community partners.

The primary objective is to improve significantly the energy efficiency of the EEC process industries, thereby helping to make them more competitive internationally and reducing the EEC's fuel import requirements. It should be noted that advances in these areas will usually lead also to pollution abatement benefits through reduced emissions and discharges of environmentally harmful substances.

Industry is a heterogeneous area with a large variety of industrial processes. The scope for reduction of costs by mass production for most energy saving technologies is therefore limited. This, and the fact that the required pay back times are mostly very short (2-3 years), hampers the possibilities for energy saving in industry.

At the present time, however, when profits are marginal in many manufacturing branches, energy saving is becoming important for the survival of an increasing number of industries. Although energy prices have decreased somewhat recently due to the economic stagnation, in the long term one may expect prices to increase again, in particular for oil and gas. Furthermore, 1992 is the year of the Internal Market, and industrial competition will increase, not only within Europe but also with the USA and Japan. Industry needs to use energy resources both as a raw material and as a fuel. Energy conservation is therefore a means of improving the efficiency and the competitiveness of industry.

The primary energy consumption of industry in the European Community is approximately 400 mtoe/year, which is more than 40 % of the Community's total primary energy consumption. The process industries account for 75 % of this amount. Among the process industries the largest energy consumers are the steel and chemical industries. Typically half of the total energy used by industry is used in the form of process heat (i.e. 200 mtoe/year) and large quantities of waste heat are dissipated. The development of heat recovery techniques is therefore very important.

The amount of process heat required in industry as a function of temperature has two peaks, one between 80°C and 200°C and a second between 800°C and 1400°C. In the whole temperature range, large quantities of waste heat are discharged which, if they can be recovered and used, can lead to large energy savings. It is the heat recovery applications to this lower peak which will be described.

The recovery and re-use of waste heat in industry formed a focal point of the programme. Within this field, a number of low temperature heat recovery technologies have been developed and studies can be divided into two main areas :

(i) development of new or improved equipment for general use in a number of industrial sectors and for various applications;

(ii) specific applications in energy-intensive industrial sectors.

2. HEAT RECOVERY EQUIPMENT

The part of the programme dealing with heat equipment focussed on heat exchangers and heat pipes, compression and absorption heat pumps, heat transformers, and organic Rankine cycle machines (ORC). To solve the problem of mismatching between energy supply and demand, energy storage technologies were also studied.

HEAT EXCHANGERS

There is strong emphasis on the development of heat exchangers, since this represents one of the largest potentials for energy conservation (e.g. 33 % of industrial energy use in France) (1).
The objectives of the EC research are to :

- improve performance and reduce the cost of existing heat exchangers;
- develop new concepts, including process intensification;
- tackle the problem of fouling and corrosion.

To that end, low temperature heat exchanger research was carried out in the following fields :

Improved performance and reduced costs. For several applications where heat exchangers form a large part of the investment costs, a better performance and a reduction in cost can make a major contribution to a breakthrough or to better competitiveness. Heat pumps represent such an application, and so heat exchanger research in this context was carried out.

Heat pipes. One merit of the heat pipe heat exchanger is modular construction which also facilitates installation. Studies concentrated on improving performance : heat transfer characteristics, vapour-liquid flow interaction and materials compatibility.

Following successful development of heat pipe modules and the associated manufacturing technology, a number of heat exchangers were constructed for demonstration in a variety of gas-gas heat recovery applications. This work involved installing units on a wool drying oven (2.3 years payback period), a hood of a paper-making machine (1.7

years payback), a laundry batch dryer (3.6 years payback), and a continuous dryer in the same field (3 years payback). Tests on these installations revealed that their performances were within 10-15 % of that predicted. Some fouling was encountered, leading to increases in pressure drop but filter packages have been designed to overcome this (2).

Work on such gas-gas heat exchangers has enabled users to have increased confidence in design procedures and lifetimes, and such exchangers are in operation in many low and medium temperature unit operations, with payback periods of 2-3 years (3-4).

Plastic heat exchanger. A heat exchanger was developed consisting of thin walled plastic tubes, to be used for a mechanical vapour recompression evaporator for viscose process liquors, see Fig. 1 (5). In 1980, it was calculated that if this work were to be successful and if the resulting technology were applied to 50 % of the viscose production in the EEC, the energy saving would be around 0.5×10^5 toe/year. An optimum design for a 50 t/hr commercial unit has been calculated to give a payback period of around 1.2 years. This was very attractive and a demonstration plant was constructed. The heat exchanger is now being commercialised.

Compact design. A different goal was set for a compact effective gas-gas heat exchanger, i.e. one in which a large amount of surface area is contained in a small volume. This is important in many "retrofit" applications where space is not available (6). Installation costs are also reduced (7). This heat exchanger is easy to manufacture, has low pressure drops and is easy to clean. The system under development can accept gases up to 400°C and could have a duty of typically 650 kW. A successful 200 kW prototype was operated under an EC Demonstration Programme for heat recovery from industrial fumes at 600°C.

Expert system. A start has been made with the development of a computer based heat exchanger "expert system" which will help to select the most appropriate heat exchanger for a specific task, taking into account a large number of variables, such as materials used, properties of fluids, surface geometry, cleaning methods, etc.

HEAT PUMPS

Heat pumps have a large field of application. This is reflected in the wide range of capacity, operating temperature levels, load, etc. The performance of a heat pump is influenced by the large number of these factors. It is therefore to be expected that a wide variety of heat pumps, mostly designed for specific applications, will be developed.

The objective of the EC research is to improve the performance and reduce the cost of existing heat pumps in order to achieve economic feasibility. Furthermore, activities are directed at developing and testing fluids for operation up to 200°C, developing adsorption heat pumps with solid/liquid combinations and improving control. The heat source for industrial heat pumps is generally waste heat. Heat pumps in industry serve mainly as a tool to transform waste heat into heat at a higher temperature level where it may be used. In the low temperature range of 50-200°C, heat recovery with heat pumps is an interesting option.

With the present state of the art, heat pumps can produce heat up to 120°C. The energy saving potential for industrial heat pumps is modest (1-2 % of the total primary energy consumption) but can be trebled if heat pumps can be developed which produce heat up to 300-400°C. The Brayton cycle offers some promise in this respect (see below).

A large part of the heat pump programme was devoted to the improvement of compressor heat pumps, and research was carried out on fluid mixtures, influence of compressor lubrication oil on the heat pump performance, oil-free compressors and heat sources. In different ways, the COP of conventional compressor heat pumps has been improved (defrosting of the evaporator, use of microprocessors for part load operation, fluid mixtures). A major part of the work was focussed on absorption heat pumps as they have the promise of being low-cost, very efficient and reliable. Work on absorption heat pumps was directed towards the development of a low-cost and reliable fluid circulation pump and of new working fluid pairs. Research on industrial heat pumps, with drying applications in the paper, milk and food industry, has been considerably expanded. The possibilities for heat pump operation above 120°C have also been explored.

The work on industrial heat pumps will be described in more detail. In the temperature range 50-150°C, several heat pump applications in industry have been investigated :

Industrial compressor heat pumps. A 330 kW_{th} ICE driven heat pump was developed which produced steam at 110°C; waste heat at 80°C served as a heat source (8, 9). The PER value is 1.5. The heat pump was demonstrated at the site were it was developed under an EC programme.

A second application for an ICE driven heat pump was investigated for grain drying in one of the most efficient working modes : simultaneous heating and cooling (10). The heat pump produces both chilled air of 4-5°C (heat source), which can be used for refrigerated storage of undried grains, and hot air up to 65°C which is used for drying. In this way the grain can be stored for 3-4 months when chilled to 4-5°C. A drier can then be utilized over a period 2-3 times longer than is typical with current practice.

The only compressor heat pump project which has the promise of producing heat at temperatures much higher than 120°C is the Brayton-cycle heat pump. A design and feasibility study was made (11). In this concept air of 1 bar is heated with waste heat to 60°C, expanded in a turbine to 0.5 bar and 2°C, reheated up to 60°C with waste heat of 90°C and compressed to 1 bar at 165°C. This heat will be used for drying in the production of milk powder. The COP of the heat pump is calculated to be 3.18. The payback time is 3-4 years. In principle this heat pump concept could be used up to temperatures of 400-500°C. An application in milk spray drying was studied, and the proposed plant layout is illustrated in Fig. 2.

Industrial absorption heat pumps. An absorption heat pump which is often considered for industrial applications is the $LiBr/H_2O$ heat pump. Two projects carried out research on this type of heat pump. Detailed work was done on the design of an absorber for a 300 kW $LiBr/H_2O$ heat pump transforming waste heat of 20-50°C into heat of 60-90°C with a generator temperature of 170°C (12). A scaled down 10 kW $LiBr/H_2O$ heat pump has been built. This led to useful operating data.

For heat pump applications in the paper industry different possibilities have been studied to produce heat of 115°C for a paper drier from its waste heat at 80 - 90°C (13). A comparison of the possible systems was made regarding technical feasibility, energy saving potential and costs. The best solution was found to be a two stage $LiBr/H_2O$ system with the evaporator at 50°C, the condenser at 115°C and the temperature of the driving steam at 155°C. A 1 MW prototype was constructed.

The search for new working fluid pairs was a focal point in the Second Programme, and several promising working fluid pairs such as TFE/NMP, R123a/E181, R22/DTG and $NH_3/LiNO_3-H_2O$ have been identified and characterized (14). Another objective was to find working fluid pairs for absorption heat pumps which are able to upgrade industrial waste heat to temperatures higher than 120°C (15). Trifluorethanol/Quinoline turned out to be a suitable working fluid pair, which compares favorably with $LiBr/H_2O$ as it achieves higher PER values and does not have crystallization problems.

For absorption heat pumps, chemical decomposition of the working fluid pair is often the bottleneck for achieving high temperature heat production. Here the absorption heat pump is at a disadvantage as the temperature of the produced heat is considerably lower than the temperature of the generator which has the highest temperature in the circuit. This is not the case with a heat transformer where medium temperature heat is given to the generator and evaporator to produce high temperature heat at the absorber. The fraction of medium temperature heat which is transformed into high temperature heat depends on the temperature difference. Here heat is produced at the highest temperature in the circuit and heat transformers are thus intrinsically better suited for achieving high temperatures than absorption heat pumps. They have the additional advantage that, apart from the waste heat input, no extra heat is needed.

A 10 kW transformer with NH_3/H_2O as a working fluid pair was developed for conversion of industrial waste heat at 100°C into steam of 135°C (3 bar) without further need of fuel or power (16). The amount of heat at 135°C delivered by the absorber was about 40 % of the waste heat at around 100°C given to the generator and evaporator. Disadvantages of the NH_3/H_2O fluid pair are the high pressures involved, which lead to more expensive tubing, and the toxicity of the NH_3.

At present, work is continuing on the identification of new fluid pairs which can meet the demands of high temperature operation. Research is also being carried out on practical new cycles such as heat transformers.

ORGANIC RANKINE CYCLE MACHINES

If recovered waste heat cannot be used for heating purposes in the factory, it may be transformed into electricity which can be transported more easily over long distances. This can be done with organic Rankine cycle machines (ORC). The ORC concept has the potential for a broad diversity of applications ranging from paper mill cooling water effluent at 60°C, to glass furnace exhaust gases at over 400°C. However, two industries have the greatest potential application for low temperature waste heat recovery systems, the petroleum industry and the chemical industry.

The following ORC machine studies for low temperature heat recovery have been carried out :

The recovery of industrial waste heat between 200 and 400°C with an ORC machine cycle was studied both theoretically and experimentally (17). The objective was to identify and specify the properties of working fluids which are, at these temperatures, chemically stable and which are low-cost to make the ORC machine economically attractive. The compatibility of the selected working fluids with the construction materials is detailed in Table 1. The results of the tests indicated that screw expanders can be used up to about 100 kW with R114 containing 5 % lubricant and up to 190°C (if a degradation of 25 % is allowed within one year). A cost sensitivity analysis was made for six ORC plants differing in the working fluid (Fluorinol 85, toluene, R114), the number of expansion stages of the turbine, the addition or omission of a recuperator, the efficiency (10.9 % to 27.5 %) and the energy recovering rate (696 to 1760 kWe). The payback period for these plants was calculated to be between 2 and 10 years. The main conclusion was that less sophisticated systems receive a higher economic benefit than systems with a high energetic efficiency. A typical low temperature ORC application in industry was also investigated. This was a hot water source at 160°C with working fluid R114 and outputs of 500, 1000 or 5000 kWel. The payback period for this plant also ranged between 2 and 10 years.

Exhaust gases of ceramic tunnel ovens leave at temperatures of 220°C and large quantities of heat can be recovered. Part of the heat was transformed into electricity with an organic Rankine cycle using tetrachloroethylene as the working fluid and the electricity was used for the power requirements of the furnace. The payback period was 3 years (18).

The possibility was assessed of building an economically attractive 100 kW ORC unit, with a heat source between 200 and 400°C. This unit could be used for industrial waste heat recovery from gas turbines and Diesel engine exhaust gases. O-dichlorobenzene was used as the working fluid since it possesses a good mix of properties (good cycle efficiency, thermal stability, general technical acceptability with respect to toxicity, flammability, etc.). The study shows that for 7000 hours/year of operation, the payback period was 5.5 years. If, in addition, use is made of the hot water obtained from the condenser for 4500 hours/year, the payback period was reduced to 2.75 years (19).

The general conclusion was that as long as electricity prices are low, ORC engines have a long payback period of about five years or longer although payback periods of two years have been stipulated (17, 20). The expenses for labour and material have increased during the last decade by a higher rate than those for electric power, so that if this trend does not change in the next years, one finds that the economy of such processes will not improve in future. Any economic assessment should also include the working fluid lifetime, corrosivity, toxicity, flammability, etc.

Some, however, expect that in a decade, electricity prices would rise considerably and that at that time ORC machines would become a profitable investment. Therefore, parts of the world where electricity costs will become high are expected to be the first markets for this technology.

THERMAL ENERGY STORAGE

Research on heat storage in the First Programme was of an exploratory nature, where a large number of compounds were tested on their suitability as heat storage materials (latent-, sensible-, and chemical heat storage) in different temperature ranges (-50°C to +450°C) (21 to 23). In this programme, the technical feasibility of latent heat and salt solution heat storage systems at 50-100°C has been demonstrated.

Economic feasibility of these systems depends on factors such as the cost of the storage system, the cost of the available energy and the number of storage cycles per year.

It was established that latent heat storage is much more expensive than simple hot water storage and that the energy density of these systems at 60-70°C is only two times higher. In most cases, hot water storage is therefore lower-priced and adequate.

A study which explored the heat storage opportunities in industry identified a limited number of applications which would lead to energy savings amounting to 1 % of the energy used for process heat. Most suitable were steam accumulators and regenerators. No industrial processes were feasible for thermo-chemical storage systems, mainly because of the rapid heat demand fluctuations which require too high a power density of the chemical storage system (24).

The greatest barriers to the implementation of thermal storage are economic; the systems have excessive payback periods or they require excessive capital outlay which will prevent other capital projects from being implemented. There is often also a conflict between expenditure on production equipment and expenditure on services, since many firms will give preference to production equipment which will give a tangible return.

3. APPLICATIONS IN INDUSTRIAL SECTORS

Energy saving by improved low temperature heat recovery may be applied in a large variety of industrial processes. Research carried out in the part of the programme on recovery in industrial processes shows that significant energy savings can be achieved by recovery and re-use of heat. The most interesting and significant results are given below :

Textile industry. A survey was made on the possibilities for energy savings in dying fibres and tissues. It was established that energy savings could be realized by heat recovery from waste water (with payback times of 12 to 35 months), from air released from dryers (with a payback time of 3 years) and from exhaust gases (25).

A synthetic fibre drying oven was used to compare the performance of a rotating regenerator and a thermosyphon heat exchanger, both used for process air preheating. Results indicated that it is not possible to propose that one type is preferable to the other : the heat wheel created more problems in respect of maintaining airflow but it had a higher efficiency. For neither of the exchangers will fouling be a problem (26).

Food industry. The preparation of foodstuffs involves many different kinds of processes, a large number of which take the form of either heating or refrigeration. This industry is noted for its intensive use of water and steam, and the resulting large quantities of effluent. It is in general easier to find uses for waste heat within such industries.

A survey in the food industry was made to detect possibilities of energy savings in this manufacturing branch. A major recommendation resulting from this survey was that more effort should be devoted to demonstrating to the food industry the value of carrying out energy audits to identify areas of wasted energy. A need to develop a heat pump to use low grade heat to produce hot water at 100°C or low pressure steam was also identified (27).

A study carried out in bakeries indicated that heat recovery from the flue gases of continuous ovens will result in considerable energy savings with a payback period of 1.5-2.5 years. Heat recovery from flue gases of the steam and central heating boilers result in a payback period of 1.1-1.5 years (28).

Another project involved gas-gas heat recovery on a spray dryer handling skim milk powder. Of particular importance in this application was the need to overcome losses in efficiency and increased pressure drop due to the combined effects of condensation and fouling on the heat exchanger surfaces. Commercially available plate and heat pipe heat exchangers were tested on the spray dryer, and it was found that the heat pipe unit, when combined with upstream filtration, could achieve paybacks of about 2.5 years. However, the thermal efficiency of the heat exchanger was low (less than 50 %) due to the necessity to employ wide fin spacing due to fouling (29).

Cooling of milk on farms for storage requires refrigeration plant. In order to save electrical energy, a simple heat exchanger which precools the milk from 35°C to 20°C was successfully applied. Using a readily cleanable stainless steel and PVC assembly, boiler feedwater preheating is an additional benefit. The system resulted in an energy saving of 34 %, and would have a payback period of less than 3 years in all except the smallest dairy farms in most Community countries (30). Use of a heat pump, chilling the milk to 4°C while heating washing water from 11° to 60°C was cost-effective, the more acceptable paybacks (less than 2 years) occuring as the size of the dairy herd exceeded 40 cows, or a milk production of 200 m^3/year.

Soya beans are the basis of popular edible oils, and hydrogenation, an exothermic reaction, forms part of the oil production process. It was demonstrated that the reaction heat may be recovered via a heat exchanger to produce hot water or low pressure steam at 100°C. On a 60 tonne/day plant, a rate of return of 20 % is anticipated (31).

Heat recovery possibilities in breweries have been studied and a detailed investigation of three breweries was carried out. Most of the heat is required at temperatures below 150°C. Possibilities for energy recovery in the malt and wort production have been identified (32).

Iron industry. The possibilities for heat recovery from a cast iron melting furnace were studied. A cupola furnace which produces 40 tonne steel per hour, produces 25 000 Nm^3/h of flue gases containing 16 - 22 % CO. Taking into account the heat losses at the chimmey where fumes are cooled to 200°C the recoverable heat is about 12.4 x 10^6 kcal/h which consists mainly of combustion heat in CO. This study showed that the heat can be used for the production of electricity, which will be used in the plant. Recovery of 87 kWh of electricity per tonne of steel produced is possible. A steam turbine driven generator of 3.5 MW will be required. Initial cost estimates indicate that the electricity produced by this method will be less expensive than if produced by a conventional generating station. A further advantage is that the CO is burnt providing an environmental benefit (33).

Further details of the projects described in this article can be found in (34 to 43).

4. FUTURE PROGRAMME

In spite of intensive rationalization efforts undertaken in the last decade, there is still an important energy saving potential in all branches of industry, particularly in the energy intensive ones.

For the next programme, the European Commission intends to develop an energy conservation R&D strategy to ensure to an even greater degree that the next R&D programme complies with the needs and possibilities of industry. A vast number of topics can be covered by strategic R&D, but the Commission cannot be involved in all of them. Therefore, it is necessary to identify priority items in which the Commission can play a useful role. The most effective way to do this is to talk to industry, and then to universities. Interviews show that industry is very interested by the Commission's intention to coordinate these activities.

Future activities identified may be directed at ensuring a Community capability in manufacturing cost effective industrial compression and absorption heat-pumps. For heat exchangers, improvements to existing heat exchangers and the problem of fouling are seen as high priorities while new concepts, including process intensification, are considered important.

5. CONCLUSIONS.

It is estimated that ultimate use of new Energy Conservation technologies, could in the long run save up to 20 % of the current industrial energy consumption (10000 MECU/year).

It is believed that continuing work in the area of heat recovery is needed to help achieve these energy savings. Heat exchangers and heat pump systems are the most attractive options in the short to medium term. They can also contribute to industrial competitiveness and pollution abatement.

Table 1. Materials compatibility and stability limits of selected fluids (MBB)

Fluid	Stability limit °C	Compatible materials	Incompatible materials
R114	200	Steel, Copper, Asbestos and paper board with solvent resistant glue, Aluminium, Nickel	Jointing material, like fat, wax, resin or natural rubber
Fluorinol 85	290 - 330	Steel of low carbon content Stainless steel	Copper * Aluminium *
Toluene	420	Steel, Copper, solvent resistant gaskets of Asbestos and paper board	Plastics, colours (Toluene is a good solvent)

* From manufacturer's information, Fluorinol 85 is compatible with Copper and Aluminium at condenser-temperature.

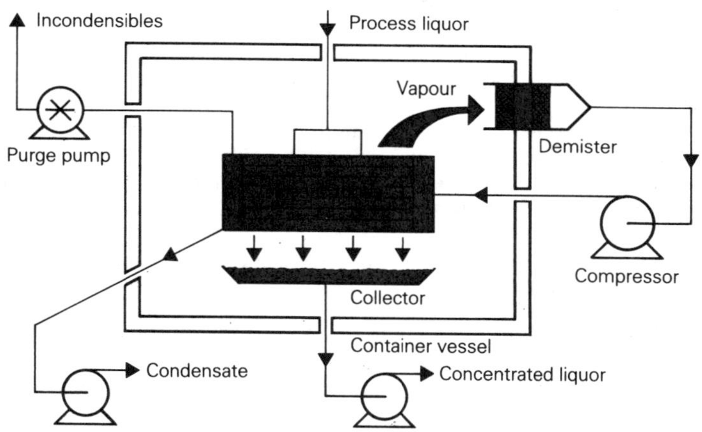

Fig 1. Essential features of mechanical vapour recompression evaporator for viscose process liquors (Courtaulds)

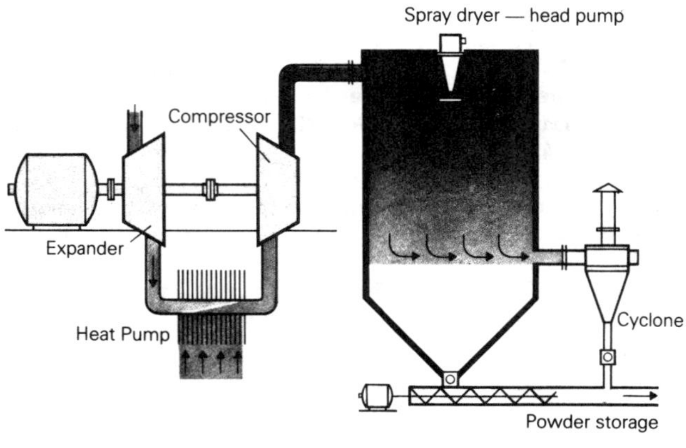

Fig 2. Brayton cycle heat pump on spray dryer (CEM)

REFERENCES

1. R. Dumon, Les échangeurs de chaleur au présent et au futur, Energie Plus, 61, 27 (1987)
2. M.J. Davies and G.H. Chaffey, Development and demonstration of improved gas to gas heat pipe heat exchangers for the recovery of residual heat, CEC, report EUR 7127 EN (1981)
3. M. Groll, N. Nguyen-Chi and H. Kraehling, Heat recovery units using reflux heat pipes as components, CEC, report EUR 7006 (1981)
4. M. Groll, D. Heine and Th. Spendel, Heat recovery units employing reflux heat pipes as components, CEC, report EUR 9166 EN (1984)
5. R. Thornton, The evaporation of viscose process liquors, CEC, report EUR 9403 EN (1984)
6. A. Grehier, C. Raimbault, A. Rojey, C. Busson, B. Chlique and J. Dreuilhe, Echangeur compact gaz-gaz, CEC, report EUR 9104 FR (1984)
7. C. Ramshaw, Process intensification: a game for n players, Chem. Engr, 416, 30 (1985)
8. D.B.A. MacMichael and D.A. Reay, Feasibility and design study of a gas engine driven high temperature industrial heat pump, CEC, report EUR 6262 EN (1979)
9. V.A. Eustace and S.J. Smith, Industrial applications of high temperature gas engine driven heat pumps, CEC, report EUR 8860 EN (1984)
10. M.B. Cunney et al, Application of engine-driven heat pumps to grain drying with refrigerated storage, CEC, report EUR 10303 EN (1985)
11. J.P. Flaux, Pompe à chaleur industrielle à cycle Brayton haute température, CEC, report EUR 9849 FR (1985)
12. T. Heppenstall, A theoretical and experimental investigation of absorption cycle heat pumps for industrial processes, CEC, report EUR 10806 EN (1986)
13. W. Friedel et al, Heat pumps for heat recovery from paper dryers, producing process steam from the dryer exhaust air, CEC, report EUR 10553 EN (1986)
14. H. Bokelmann and H.J. Ehmke, Working fluids for sorption heat pumps, CEC, report EUR 10725 (1986)
15. J. Berghmans, Development of an absorption heat pump for industrial application, CEC, report EUR 10432 (1986)
16. J. Engelhard, Development of heat transformer which produces process steam at 130°C, CEC, report EUR 10807 DE (1986)
17. G. Huppmann, Nutzung Industrieller Abwärme durch ORS-Systeme, CEC, report EUR 9271 DE (1984)
18. Macchi et al, Heat recovery by organic Rankine cycle in ceramic firing ovens, CEC, report EUR 7642 (1982)
19. A. Angelino, M. Gaia and E. Macchi, Design, construction and testing of a hermetically sealed 100 kW organic Rankine cycle engine for medium temperature (200 + 400°C) heat recovery, CEC, report EUR 10324 EN (1985)
20. ORC-HP Technology, Int. VDI-Seminar, VDI Verlag (1984)
21. P. Eckerlin et al, R&D of systems for thermal energy storage in the temperature range from -25°C to 150°C, CEC, report EUR 6936 (1980)
22. M.A. Bell and I.E. Smith, Thermal energy storage using saturated salt solutions, Energy, 5, 1085 (1980)

23. P.W. O'Callaghan et al, Thermal energy storage systems, CEC, report EUR 7266 EN (1981)
24. D.T. Baldwin et al, Energy cascading combined with thermal energy storage in industry, CEC, report EUR 8904 EN (1984)
25. J. Laneres, Economy and energy saving in the textile industry, CEC, report EUR 7302 (1981)
26. D.B.A. MacMichael, D.A. Reay and B.L. Forster, Comparative study of rotating regenerators and heat pipe heat exchangers, CEC, report EUR 6792 EN (1980)
27. W.E. Whitman et al, Energy saving opportunities in the UK food industry, CEC, report EUR 7073 EN (1981)
28. L. de Vries et al, Energiebesparing in de bakkerij door een doelmatiger gebruik van energie en door terugwinning van afvalwarmte, CEC, report EUR 10087 NL (1986)
29. L.A. Jansen et al, Recovery of heat from exchaust air of spray driers in the dairy industry, CEC, report EUR 7576 (1981)
30. J. Ubbels et al, The saving of energy when cooling milk and heating water on farms, CEC, report EUR 6915 (1980)
31. T.L. Ong, Recovery of residual heat in the extraction of oil seeds and in the hydrogenation of edible oils and fats, CEC, report EUR 6837 (1980)
32. T.S. Kampffmeyer, Research on energy saving in a brewery, CEC, report EUR 6666, Reidel (1979)
33. A. Calabro' and M. Misschiatti, Energy recovery from cast iron melting furnaces, CEC, report EUR 10524 IT (1986)
34. A.S. Strub and H. Ehringer, New ways to save energy, Proceedings of the International Seminar, Brussels, 23-25 October 1979. D. Reidel (1980)
35. A.S. Strub and H. Ehringer, Energy Conservation in Industry, Proceedings of the International Seminar, Düsseldorf, 13-15 February 1984. VDI (1984)
36. P. Zegers, The Community's Energy R&D Programme - Energy Conservation : Survey of Results (1975-1979), 2nd Edn. CEC, report EUR 7389 EN (1982)
37. H. Ehringer, G. Hoyaux, P.A. Pilavachi and P. Zegers, The Community's Energy R&D Programme - Energy Conservation : Survey of Results (1979-1983), 2nd Edn. CEC, report EUR 8661 EN (1986)
38. H. Ehringer, G. Hoyaux and P.A. Pilavachi, Energy conservation in industry - combustion, heat recovery and Rankine cycle machines. Proceedings of the Contractors' Meetings. D. Reidel (1983)
39. H. Ehringer, G. Hoyaux and P.A. Pilavachi, Energy conservation in industry - Applications and techniques. Proceedings of the Contractors' Meetings. D. Reidel (1983)
40. D.A. Reay, Heat Recovery - Research and Development within the European Community, Heat Recovery Systems, 2, 419 (1982)
41. P. Zegers and J.A. Knobbout, An overview of work on industrial and domestic heat pumps in the energy R&D Programme of the European Community, Int.Symp. Ind. Appl. of Heat Pumps, Warwick, 243 (1982)
42. P. Zegers, Results of the Heat pump R&D Programme of the European Community, 2nd Int. Symp. The Large Scale Appl. of Heat Pumps, York, 311 (1984)
43. P.A. Pilavachi, Energy Conservation R&D in industry, Heat Recovery Systems & CHP, 7, 329 (1987)

SESSION IV : INDUSTRIAL PLANT – PROCESS CONTROL AND OPTIMIZATION

Control and optimization of processes

An unconventional energy recycling project

The optimized process control of an ethylene plant

Microprocessor system and digital regulation loops for increasing cowpers energy savings

CONTROL AND OPTIMIZATION OF PROCESSES

Boris KALITVENTZEFF
University of Liege
Royal Military Academy, Belgium.

1. INTRODUCTION

Improving performances in industrial processes is a constant concern : the problem comes up at the plant design stage, again if an extension or a modification of an existing process is being envisaged; finally when choosing the operating conditions and the control of the plant in view of the production required : the economic conditions at a given time poses the question in constantly changing terms. Over the last few years, process control has evolved : it started with single analogic loops, the set points of which were modified manually, and is now evolving towards direct digital control where the process computer can decouple the variables and adjust the independent parameters to the operating conditions.

A hierarchical command system, based on a group of microcomputers or microprocessors supervised by a central computer which distributes the set points can result in a constant optimization of working conditions.

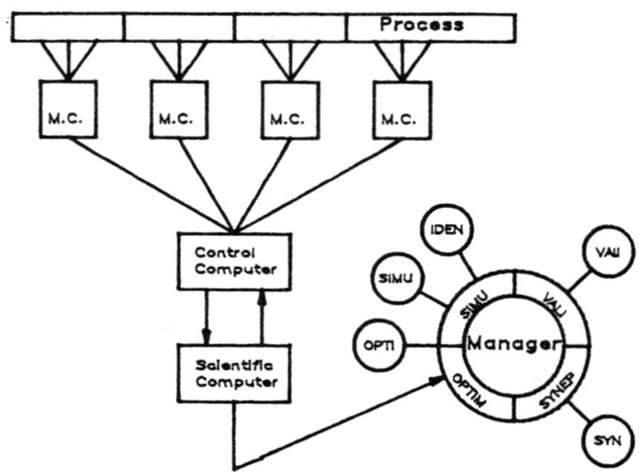

Figure 1 : Hierarchical command system

But there is still a long way to go before this ideal situation is generalized : current data processing equipment is not yet able to economically solve the problem of the optimal operation of a flexible process by applying systematic analytical methods.

2. OPTIMIZATION AT WHAT LEVEL?

Industrial plants are composed of three interacting sub-systems (figure 2) : the operating units, which process raw materials into finished products and waste, the heat exchange and recovery network, and the sub-set handling the supply of utilities (steam, water, electricity, air). The objective is to economically exploit the group of production equipments. But the system is sufficiently complex that often optimization studies only cover a fraction of the whole process.

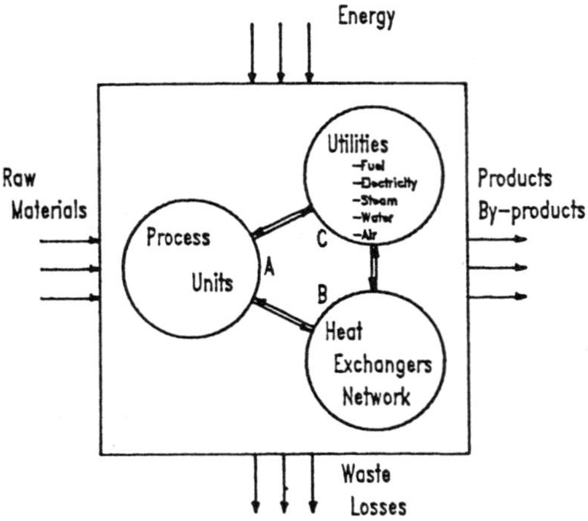

	Design	Retrofit	Control & Management
Process Unit	++		++
Exchangers Network	++	+	+
Utilities	++	+	+
Total Plant	+		

Table 1. : Available literature

Figure 2 : Interacting sub-systems

Table 1 gives a classification of the problems to be solved, and indicates whether solutions have been proposed in technical literature.

2.1 <u>Designing the process</u>

The techniques for rational design of (A) operating units taken on their own, (B) the exchange network or (C) the utility supplies, are discussed in technical literature. However, this type of analytical approach does not make it possible to find an optimal configuration for the whole process : a logical and systematic approach would take into account the interactions between sub-systems. Certain studies consider the sets A+B together (for example simultaneously optimizing nominal operating conditions in a distillation train and the heat exchange network), while considering the cost of utilities as fixed. In fact, these costs depend on demand, and are not independent : a large demand for high-pressure steam on a site can drastically bring down the marginal cost of low pressure steam for example. Similarly, the simultaneous analysis of sub-systems B+C consists in taking a static image of the process and the utility demand (both quantitative - level of temperature and pressure - and qualitative). In fact, these parameters can often be modified significantly by altering operating conditions or the sequence of production operations, thus favouring synergy and affect by the need for utilities considerably.

On developing a process, it is advisable to provide sufficient flexibility in the way it operates equations so that it can be optimized under various market conditions. In particular, the choice of the independent variables in regulator loops must remain pertinent for a large range of operating conditions.

It does seem necessary therefore to completely integrate sets A+B+C, but the optimization methods currently available, which will be described below, are not able to resolve all of the major problems corresponding to more complex processes at one time.

2.2 <u>Upgrading existing plants</u>

Studying extensions or transformations of existing processes is done less frequently. They do, however, have a significant practical interest and introduce many constraints due to conservation and re-use of existing equipments. Moreover, this is not possible without thorough knowledge of the state of the process which is to be modified. An appropriate means of handling the measurements and a methodology are given below.

2.3 Optimizing production

Finally, optimizing all of the operations from day to day, given the various production targets, technical (defective equipment, limited availability supply of certain supplies), climatic or economic conditions is of increasing concern to industrialists. The following questions must be answered to design processes which offer more flexibility and are sufficiently adaptable :

- to what extent is the process flexible?
- What command variables, corresponding to the set-points in the control loops, should be used to pilot the process ?
- What is the best choice of variables manipulated by the controllers, and how should they be associated with the regulated variables so as to decouple interactions among variables as much as possible ?
- How should the optimal value of the set-points be determined?

The answer to these questions also requires an in-depth analysis of the process, which should be undertaken in part starting at the plant design stage, but certainly during operation.

Using simulation software makes it possible to assess the cost of the various production policies which can be envisaged. The static simulation model is the basis of an optimization study and a fundamental factor is that the mathematical model must be adjusted to realistically approximate the usual operating conditions. After all, optimization will seek an optimum for the model, but this will only correspond to the optimum process if the model is adequate.

Finally, by comparing a series of operating conditions, the model will determine the sensitivity matrix used to look for the optimal association of control and dependent variables.

To optimize the design, operations and control of a process, there is first a need for softwares which will enable the engineer to go beyond the static design of a flow sheet, or even rating simulation.

3. STAGES OF THE ANALYSIS. THE BELSIM METHODOLOGY

Before optimizing the way a process works, a thorough knowledge of the plant and its operating constraints is mandatory. The methodology proposed to study the process is illustrated below (figure 3).

The first stage is to take measurements. These measurements may be redundant, insufficient or erroneous, because they do not satisfy the mass and energy balance constraints.

The validation stage consists of reconciliating the measurements with the balance equations to obtain a coherent set of data and to calculate state variables of the system which are not measured directly, but which can be calculated using the available measurements. This stage cannot be neglected under any circumstances because the optimization will be meaningless if the initial state of the system has not been determined objectively.

To validate the measurements, the engineer will select in the program library those modules which correspond to the mass and energy balances of each operating units in his process. He will also describe how the units are connected. This information will enable the program to generate the constraint equations of the problem.

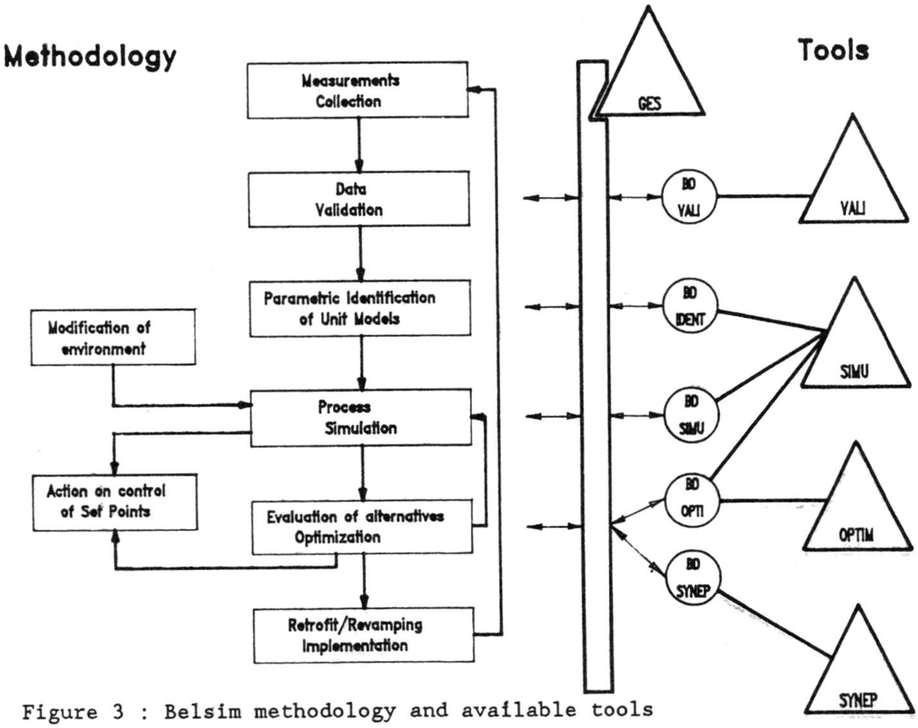

Figure 3 : Belsim methodology and available tools

Once the list of measured variables has been defined, a mathematical analysis of the problem occurrence matrix will show whether measurements are redundant, locally insufficient or just sufficient (Ph. Joris and B. Kalitventzeff, 1987). Based on the balance equations, additional measurements may be proposed. At this stage of the methodology, the engineer can check whether the entire set of measurements is sufficient to define the state of the system and therefore to determine the number of degrees of freedom. Validation is the solution of a constrained optimization problem. The equation to be minimized is quadratic and subject to the balance constraints. The problem can be set down as follows :

$$\min\ (Y-Ymes)^t\ .W.\ (Y-Ymes)$$
$$\text{subject to } F\ (X,Y) = 0$$

where X = array of non-measured variables
Y = array of measured variables
Ymes = array of measured values
F (X,Y) = constraint equations
W = weight matrix to take into consideration the accuracy of the measurements, e.g. Wii = 1/variance associated with measurement i.

Knowing a consistent state of the process, the engineer can already assess the quality of the plant operation and can take decisions without going on to further stages.

The unit simulation parameters will be based on the knowledge of a coherent state given by the validation, but the engineer must choose the simulation models which describe the various unit operations. He will also choose the thermodynamic models which best represent the behaviour of mixtures.

Based on the validation program, and the unit models selection, the engineer will determine the command variables and go on to the next stage : simulating the entire process or parts thereof. The engineer can check that the operating state and the behaviour of the process is properly reproduced by the simulator and thus check the validity of the unit model selection.

With this simulation, one already examines the plant's response to change in command variables, variations in utility supplies, etc. For example, he can calculate the relative gains or sensitivities of controlled variables in relation to the manipulated variables in the system.

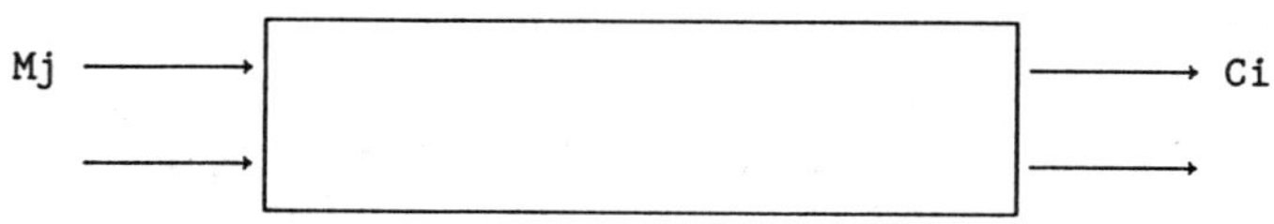

Ci = controlled variables Mj= dependent variables

The relative gain Lij is defined as the ratio of the open loop gain to the closed loop gain :

$$Lij = Jij * Hij$$

where $Jij = dCi/dMj$ and H is the transposed inverse of matrix J

The simulation model can be used for on line control/command, and is thus more than a decision-making aid.

The last stage in the methodology tackles synthesis and optimization. It can include simple assessments of alternatives using successive simulations, any energy analysis of the heat exchange network, or non-linear optimization of a cost function under constraints. This last stage is the most expensive and the least systematic. It entails sophisticated mathematic methods and above all undeniable competence in process engineering.

Data processing tools are required for the various stages in the methodology. We apply this methodology using BELSIM (User's Manual 1987) software which includes various integrated modules : a validation module, a module for fitting parameters and for simulation, a costing module, an energy integration module and an optimization module.

It is easy to understand that often one may have to go back to gathering measurements, choosing models or to process simulation. The various tools used in each stage must absolutely be integrated into a "manager" program and all of the information gathered at all levels (measurements - validation - identification - simulation - synthesis) must be organized into a data base. The PDB (process data base) includes the data and the results of the various stages of the methodology. Interactive programs allow to update the PDB and to estimate physical properties of any stream defined in the PDB.

4. OPTIMIZATION METHODS

The general optimization problem takes the following form :

$$\min F(x, y) \text{ with } \begin{array}{l} h(x, y) = 0 \\ g(x, y) < 0 \\ x_{min} < x < x_{max} \\ y = 0 \text{ or } 1 \end{array} \quad (1)$$

where y represents the set of discrete decision variables and x the set of system state variables. Before looking into the methods for solving this very general, mixed integer non linear programming problem (MINLP), we will first consider a more limited problem obtained by fixing the value of the integer variables.

In the case of process management, the objective function to be minimized represents the plant operating costs, which sums up the various criteria used to optimize each of the sub-systems (cost of utilities, raw materials, energy, etc.). A complementary term for investments can be included for the design problem. The equality constraints are equations describing the process (balance and rate) and the specification constraints.

The inequality constraints set limits on the process operating range : they are physical constraints (positive flowrates) or technological constraints (range of tolerable flowrates in a pump or a turbine, temperature or pressure limits in the equipments) or numerical constraints (arguments of a logarithmic equation greater than 0). These constraints may also represent safety limits of the process control system. The market or environment constraints (purity of products, pollution standards, etc.) define another type of inequality constraints. These equations are of capital importance because the optimum is often located on this type of constraint.

Three strategies can be envisaged to solve this problem. We will discuss their respective advantages and disadvantages. We can already see that there is no universal strategy, and the choice of the appropriate method in resolving the problem will depend from the type and size of that problem.

4.1 " Black Box" Optimization

This strategy is represented by the following diagram (figure 4)

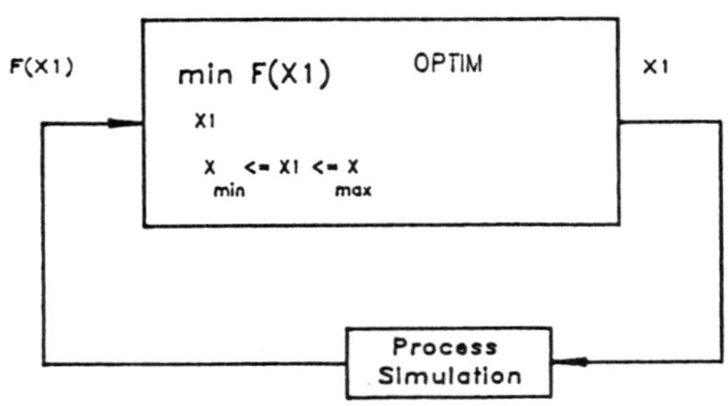

Figure 4

"Black Box" Optimization.

where X1 represents a set of independent variables chosen by the user. The simulator uses these variables to solve the process equations, and computes the objective function value. It acts as a "black box" for the optimizer, who is faced with a problem of the form :

$$\min F(X1) \text{ subject to } Xmin < X1 < Xmax$$

This problem, which does not have any equality constraints, can be treated using classical methods (complex, conjugate gradient, etc.)

This is a robust approach, but it is very expensive in computation ressources : each set of control variables and each perturbation of these requires a complete simulation of the process for the calculation of the gradient. It can only handle inequality constraints which are explicitly defined as boundaries on X1 independent variables. It is therefore limited to cases where the number of independent variables is small (at most about ten), where the process is easy to simulate and the optimum is not subject to constraints other than Xmin and Xmax boundaries. The choice of independent variables is particularly important in this case; one should define the operating point by choosing the specifications which are the most sensitive with regard to the objective function, or those which are needed to define inequality constraints. This method can be used with most of the simulators currently available.

4.2 <u>Sequential modular approach</u>

Processes with recycles can be simulated iteratively . In the sequential modular approach, a subset of "torn" variables X_2 is selected and initialised, such that each piece of equipment can be calculated in sequence (unit inputs being either torn variables, output of already calculated units or system inputs). Calculating the sequence provides new values x_2' of torn variables. Thus matching x_2 and x_2' can be handled in the optimization by introducing constraint equations $h(x)=0$ to represent the tear equations $x_2 - x'_2(x_1, x_2)=0$, x_1 being the independent variables. This strategy corresponds to the schema below :

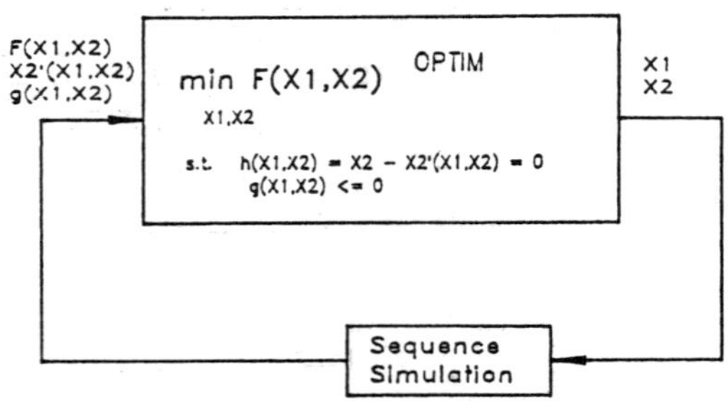

Figure 5

Sequential modular approach

The advantage of this approach is that it simultaneously handles tears and optimization. It does not require converging the system to estimate derivatives, but only by calculating a sequence. However, the method of resolution to be applied must be able to handle an optimization problem with non-linear constraints. Some available methods are :

- the "infeasible path methods" whereby the step proposed at each iteration does not necessarily meet the equality constraints. For these methods, only the solution point has a physical meaning. Successive quadratic programming methods (SQP) are in this category as are successive linear programming methods (SLP).

- In the "feasible path methods", each main iteration generates an operating point of the system. These methods mainly include the generalized reduced gradient (GRG) method and the Sargent method.

Hybrid methods have also been proposed, where some iterations to solve the equality constraints are performed once a step has been determined by the infeasible path method : (SQP partially converged). More information on these can be obtained in Biegler (1987), Lasdon (1983) and Fletcher (1981).

To illustrate this classification, the figure below shows the steps proposed by GRG and SQP for 2 starting points while solving a simple two dimensions problem. In the first case, all constraints are verified from start ; SQP steps through the infeasible domain before verifying all constraints at solution, whereas GRG remains on the equality constraint. The second case illustrates what happens with an infeasible starting point. GRG first solves the equality constraints before optimizing. SQP remains in the infeasible domain untill the solution is reached.

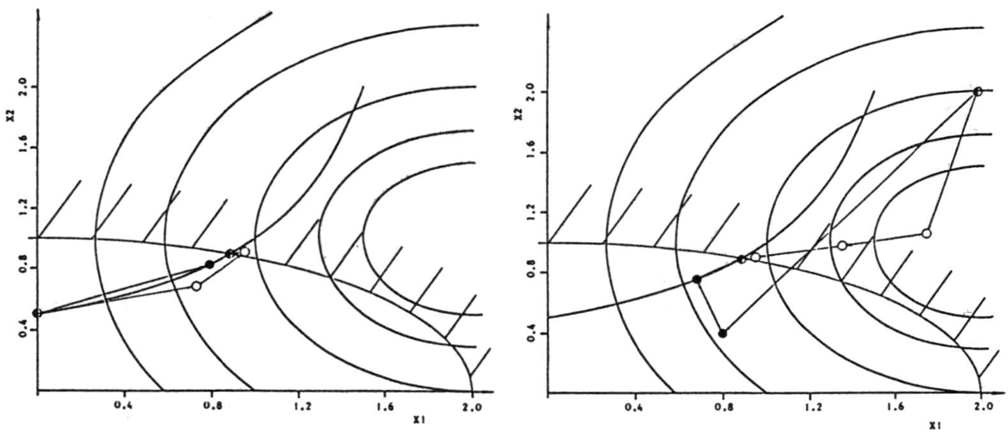

Figures 6 and 7 : ● GRG method ; ○ SQP method.

The advantage of this strategy is, as said, the simultaneous treatment of tear equations and optimization, which decreases the number of derivative calculations. However inequality constraints which do not depend explicitly from : X_1 and X_2 are difficult to handle : internal system variables which are hidden to the optimizer, are needed to evaluate them.

The optimization program would therefore have to be particularly robust, essentially for handling inequality constraints : experience shows that a solution on the boundaries is harder to find. This strategy, which is rather easily adaptable to software presenting a sequential modular structure, will preferably be applied to convex problems where number of active inequality constraints at the solution is limited.

4.3 Simultaneous resolution

This third approach consists of writing the modelling equations of all the process units as functions to be driven to zero. The large scale equation system thus obtained is represented by :

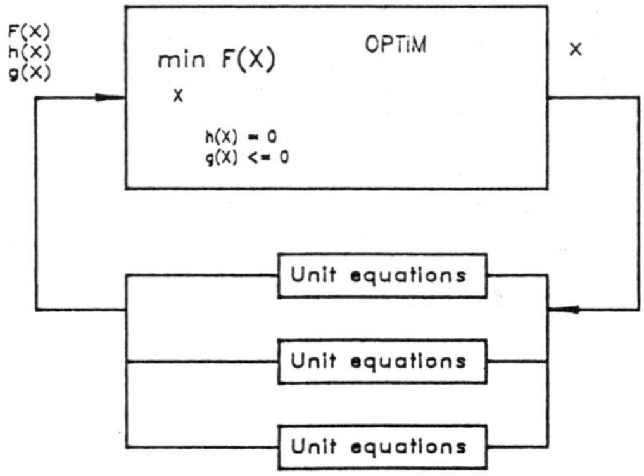

Figure 8 :

Simultaneous resolution.

Handling such a large number of equations may be difficult unless the solution method takes into account the sparsity of the problem : only a few variables appear in each equation.

To our knowledge, only MINOS (feasible path) and successive linear programming methods (infeasible path) compete. The large number of variables and equations, moreover, make this approach very sensitive to the starting point and the way functions and variables are scaled. It can be advantageously applied to large linear systems and to simulators already set up as "equation solvers". Its major advantage is the freedom in the selection of decision variables.

4.4 General remarks

Each of these strategies has its advantages and disadvantages, and a strategy that is well adapted to one problem is often not suitable for another. Particular attention should be paid to the problem formulation and the starting point for the optimization, whatever strategy is used. Indeed, a bad starting point or an inappropriate variable or equation scaling can seriously hinder the resolution. Presently available algorithms may face troubles when tackling problems where the solution lies on active inequality constraints : this is usually the case in real life situation, as shown by own experience.

4.5 Discrete variables

Finally, let us consider the problem of discrete variables. They represent yes/no decisions which come up in management problems. They introduce discontinuities in the constraints equations and in the cost function, and make the problem much more difficult to solve. The Benders decomposition theory (Geoffrion 1972) allows to break down the problem into one handling continuous variables (NLP), embedded into another one treating discrete variables. This strategy leads to one of the two following approaches. Either the discrete variables are considered as continuous and a "Branch & Bound" algorithm is applied. Or the value of the discrete variables are set, the resulting non-linear problem is solved, the equations are linearised around this solution, the solution of the resulting mixed integer linear programming problem generates new values for the discrete variables (Grossmann, 1987).

One understands easily, that solving this kind of problem is not just a routine.

5. EXAMPLE

The example given is a management problem for utility networks ; it illustrates mixed integer non linear programming. The turbines and exchangers in figure 9 operate at a given charge or power, thus the interactions between the utility network and the two other subsystems which make up the process are frozen.

Vapour can be supplied by two boilers whose operation is optional with a minimum production level. The power required can be supplied either by turbines T1 to T5 or by electric motors.

Table 2.1. : turbine characteristics

Unit	T6	T1	T2	T3	T4	T5
Yield	0.75	0.5	0.6	0.6	0.6	0.55
Power(kW)	8190	1115	690	600	261	550

Table 2.2 : heat exchangers load

Unit	E4	E1	E3
Load (kW)	1900	16000	24000

Given a cost of 1.5 units/kmol for the supply flow at 42 bar and 1 unit/kmol for the flow at 12 bar, let us consider the cost of electricity on the solution.

Table 2.3. : solution

El. cost	1.E-4	2.E-4	4.E-4
Feed 42 bar	0.741	1.207	1.690
Feed 12 bar	0.938	0.481	0.0
T1	0.0	0.0	0.4825
T2	0.0	0.249	0.249
T3	0.0	0.216	0.216
T4	0.0942	0.0953	0.0957
T5	0.217	0.217	0.219
Total Cost	2.291	2.514	2.536
Constraints	V1	V1	Feed 12 bar
	S3	S3	S3

Table 2.3 shows how the steam flowrate at each turbine is modified when electricity cost varies from 0.0001 to 0.0004 units. When electricity cost is low, 3 turbines remain idle. For a medium cost, using turbine T2 and T3 becomes profitable. For still higher price, all turbines are running, which increases the demand for high-pressure steam and 12 bar steam production is no more needed. At each solution, several inequality constraints are active. For the lower and medium electricity cost, they are minimum flowrates in V1 and 3 bar outlet S3. At higher cost, active constraints are minimum flowrate at S3 and minimum flowrate for 12 bar boiler (which thus becomes idle). Turbines T4 and T5 are used in each solution, since specified load of exchanger E3 sets the demand of 3 bar steam ; it is more profitable to use the steam first in turbines, than to

let it down in valve V2 to satisfy the low-pressure steam demand. Notice that steam rates for turbines T4 and T5 vary slightly, since varying rates in T1, T2, T3 modify the steam inlet conditions in T4 and T5.

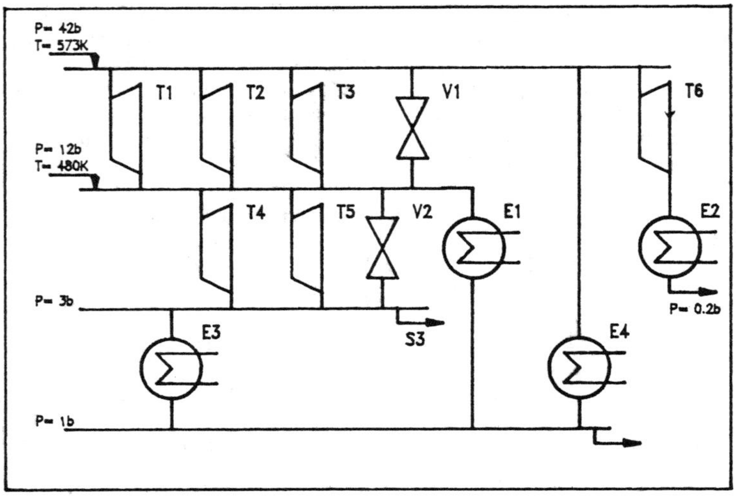

Figure 9 : Utility network

AKNOWLEDGEMENTS

The author wishes to express his thanks to his co-workers G. Heyen, Ph. Joris, F. Maréchal and J. Rennotte, for helping him in preparing this communication.

REFERENCES

(1) DURAN M.A., GROSSMANN I.E., A mixed-integer non-linear programming algorithms for process synthesis. AICHE vol 32, n 4, pp 592-606, 1986.

(2) FLETCHER R. , Practical methods of optimization ; constrained optimization. John Wiley & Sons , 1981.

(3) GEOFFRION A.M., Generalized Benders decomposition. Journal of optimization theory and applications. Vol. 10. n 4, 1972.

(4) P. JORIS, B. KALITVENTZEFF, Process Measurements Analysis and Validation. Proceedings of the CEF 87 Congress, pp. 41-46, Giardini Naxos (Sicily), April 1987.

(5) LANG Y-D, BIEGLER L.T., An unified algorithm for flowsheet optimization. Computers and Chemical Engineering. Vol. 11, n 2, 1987.

(6) LASDON L.S., WAREN A.D., Large scale non linear programming. Computers and Chemical Engineering. Vol. 7, n 5, 1983.

(7) BELSIM Users' Manual, BELSIM s.a. ed., Domaine universitaire du Sart Tilman, Université de Liège, 1987.

AN UNCONVENTIONAL ENERGY RECYCLING PROJECT

Prof.ir. H.P. VAN HEEL
Managing Director of Hoechst Holland N.V., Vlissingen

Hoechst, one of the world's largest chemical concerns, was founded in Höchst, near Frankfurt, in 1863. The turnover in 1986 was DM 38 000 million; the number of employees exceeded 180 000.

Hoechst Holland N.V., a 100% subsidiary of Hoechst, was set up in 1950. The turnover in 1986 was more than DFL 1300 million; personnel number more than 2200. The Board of Management and the Sales department are housed in Amsterdam. Hoechst Holland has two production locations: one in Weert, with about 1000 employees, where foils and decorative laminates are manufactured, and one in Vlissingen.

Hoechst Vlissingen was built on a 125 ha site, which was reclaimed in the Westerschelde river. Approx. 1000 people are employed there. About 1 million tonnes of solid goods and approx. ½ million tonnes of liquid goods are brought in annually. Electricity consumption is 1 - 1.25 x 10^9 kWh; the nearby electric power station is contracted to supply 170 MW power.

Products

- P elementary phosphorus
- H_3PO_4 phosphoric acid
- STPP sodium tripolyphosphate
- DMT dimethylterephtalate
- A.S. alkane sulphonate

Fig. 1: The products of Hoechst Vlissingen are of a very varied nature, as are the production processes.

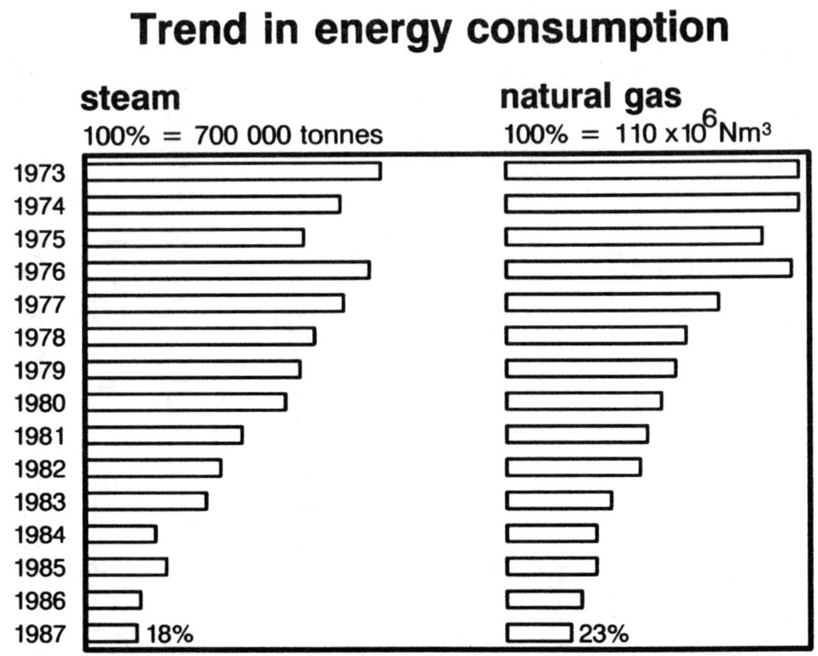

Fig. 2: In the course of the years and in spite of increased production, consumption of steam and natural gas has decreased to about 1/5, as a result of systematic energy saving. The decreased demand for steam made it necessary to shut down a Total Energy Installation (6 MW; 18 bar steam) and to put it into mothballs.

Process and energy network

$$2Ca_3(PO_4)_2 + 6SiO_2 + 10C \longrightarrow 6CaSiO_3 + 10CO + P_4$$

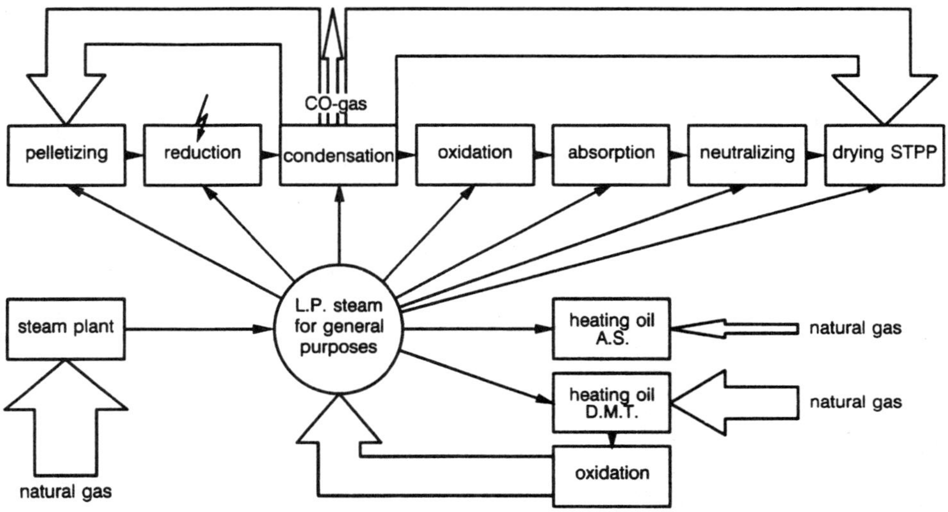

Fig. 3: Calcium phosphate (phosphate ore) is pelletized and reduced in phosphorus furnaces, using a large amount of electrical energy. The reducing agent coke combines with oxygen to form CO, which is used in the pelletizing and STPP plants for heating purposes. SiO_2 combines with calcium from the ore to form calcium silicate, a much used building material for roads and dikes. The phosphorus becomes P_2O_5 via oxydation and then, by absorption in water, it becomes H_3PO_4: phosphoric acid. Neutralisation with sodium hydroxide (NaOH) and drying in CO heated spraying towers results in sodium tripolyphosphate (STPP: $Na_5P_3O_{10}$).

Two completely separately functioning processes (alkane sulphonate, main ingredient of rinses, shampoos; DMT, dimethylterephtalate, raw material for polyester textiles and tapes) are operated at temperatures above 300° C; to this end, heating oils, brought to temperature in natural gas fired furnaces, serve as heat transfer medium.

In the DMT process, much heat is released during oxydation of the raw material paraxylene. This heat is added as LP steam to the General Purpose Low Pressure Steam Net; the output of the steam plant has fallen in the course of the years and is now at a minimum. Conversion of waste heat (from the phosphorus oxidation, for example) into LP steam is thus pointless.

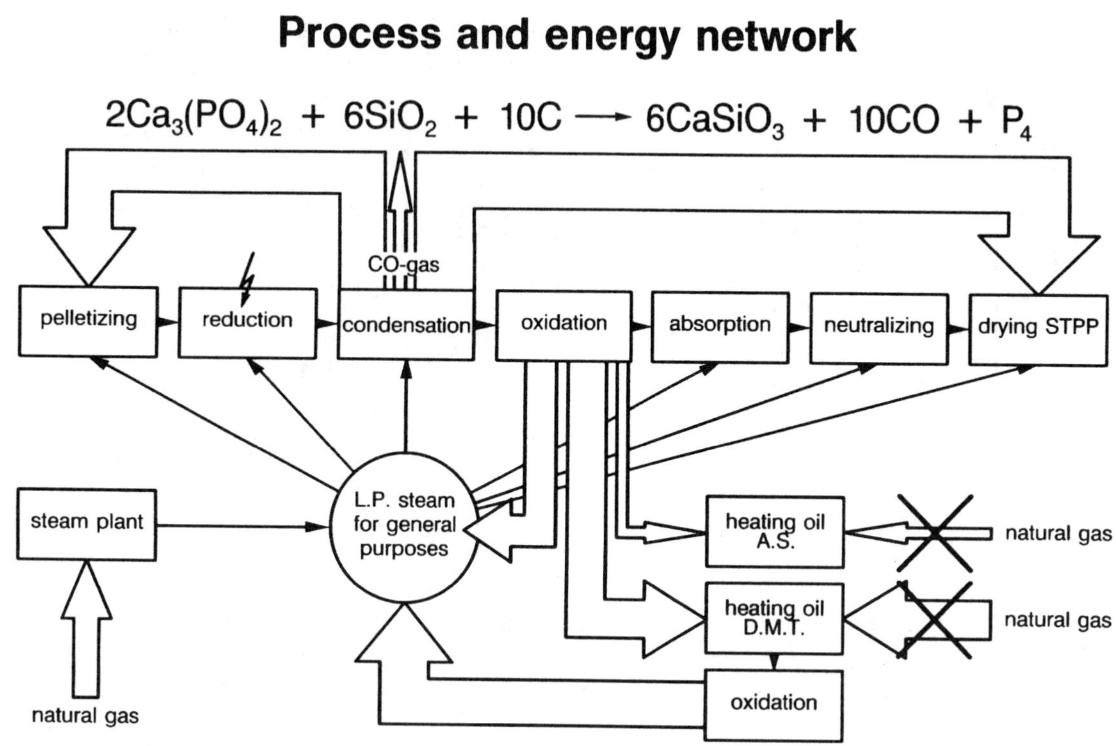

Fig. 4: The energy from the phosphorus oxidation is converted into steam of 170 bar. A new 170 bar steam piping system, which is many kilometres long, connects the phosphorus oxydation plant with the A.S. and D.M.T. plants. The natural gas valves under the heating oil furnaces will be closed. The steam plant will stand by for emergencies.
Note that the natural gas flows at A.S. and D.M.T. have disappeared, there is a reduced flow at the steam plant and there is an energy flow from the oxydation.

Acid production without recovery of waste heat

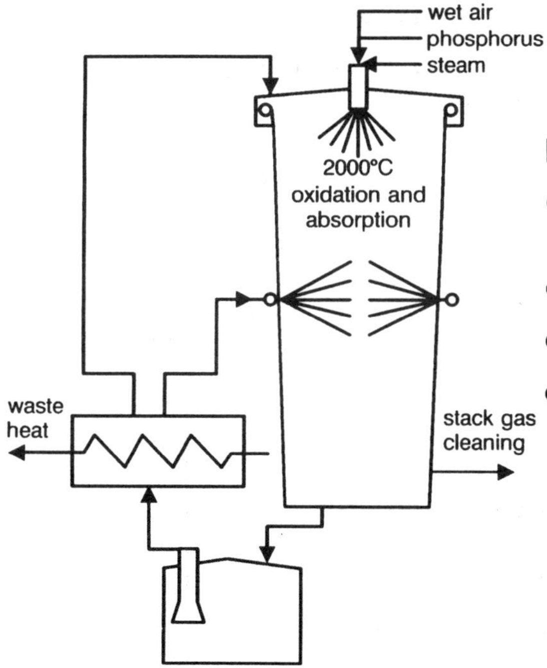

Requirements for using waste heat

- high temperature (345°C)
  high pressure (170 bar)
- stainless steel boiler
- the use of dry air for combustion
- no temperatures exceeding 420°C
  due to intercrystalline corrosion

Fig. 5: Phosphorus and steam are atomized into the top of a 20 m high tower and brought to combustion. The flame temperature is approx. 2000° C. Water is sprayed in the tower into which the phosphoric oxide (P_2O_5) is absorbed. The H_3PO_4 (phosphoric acid) thus formed circulates via a tank through a battery of cooling units, which remove the oxidation heat, and then subsequently travels along the walls of the tower, which are thus protected from the high temperatures. Recovery of the waste heat is only possible and worthwhile if the conditions are met which are shown on the right of the diagram.

Acid production <u>with</u> recovery of waste heat

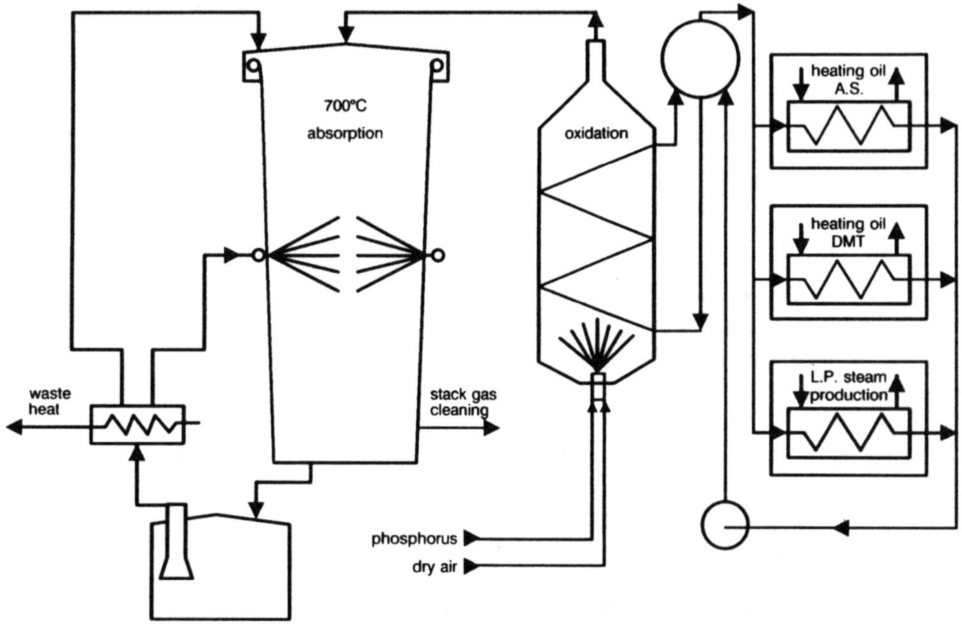

Fig. 6: Oxydation and absorption are separated. Steam of 170 bar is produced in the newly built steam boiler. The stack gases from the steam boiler are and will remain the main product and are converted to phosphoric acid in the existing acid tower. The disposal of waste heat (left) has been greatly reduced. Heating of the heating oils in the D.M.T. and A.S. plants has been switched from natural gas to steam of 170 bar. For balancing purposes an HP/LP heat exchanger has been added (below right).

RECOVERY: 5×10^{14} J/a
≈ 15 000 000 m^3 per annum natural gas

The impossibility of buffering energy in the form of CO and/or HP steam and the resulting close coupling between 10 plants operated separately from 7 control rooms, made computerized on-line communication and on-line process control necessary. Only in this way the next step on the way to further energy saving could be made: started with 115 million m^3, now via 25 million m^3 to 10 million m^3 natural gas per annum.

Specific data for process gas cooler

Dimensions	diameter	4 m
	height	17 m
	mass	31 000 kg
Material	1.4571	
Pressure	180 bar	
Temperature	380 °C	
Thermal energy	19,6 MW	
Evaporating surface	335 m^2	
Superheating surface	153 m^2	

Photo 1 - Heat exchanger being transported

Photo 2 - Heat exchanger being set up

THE OPTIMIZED PROCESS CONTROL OF AN ETHYLENE PLANT

V. KAISER, X. HURSTEL, TECHNIP France
S. BARENDREGT, PYROTEC, The Netherlands

1. CHARACTERISTICS OF AN ETHYLENE PLANT

High purity light olefins (ethylene at 99.95 %, propylene at 99.5 %) and fractions with high diolefins and aromatics content (C4 fraction with 50 % butadiene, C6/C8 fraction with 70 % aromatics) are obtained by steam cracking from paraffinic hydrocarbons.

Feedstocks for the plant are :

- gaseous : ethane, propane, butane
- liquid : naphthas, kerosene and gas oils of various origins. The density of the naphthas has a considerable impact on yields, the best for olefins being the light naphthas, density up to 700 kg/m^3.

The paraffinic hydrocarbons(ethane, propane, butane, naphthas and gasoils) are preheated (80 to 120°C) and charged with dilution steam to the vertical tube cracking furnaces (about 6 to 10 units in one plant).

The mixture is then heated to 820-860°C, under low pressure (about 2 bars), to promote the decomposition of paraffins.

The high temperature heater effluent is quenched to stop secondary reactions, thereby generating high pressure steam.

Gasoline and fuel oil are separated out in the primary fractionator. Gaseous products are cooled by direct contact with process water, and then compressed at about 36 bars.

After removal of acid gases (H_2S, CO_2) and water, the gaseous mixture is processed in the refrigerated separation train into pure components (see figure 1).

Table 1 gives a summary of product slates and energy input for various feeds.

In table 2 the breakdown of the main variable costs is presented for typical naphtha feed. It is clear that energy is a major item to be considered for the profitability of such a plant.

2. THE ADVANCED PLANT MANAGEMENT CONCEPT

The APMS (Advanced Plant Management System) package performs supervisory tasks on a level above that of advanced process control.

The important features of Advanced Plant Management are :

. It assists plant operating personnel in running the plant at optimized conditions. Plant operating personnel are continuously provided with information on the current status of the operation and its performance, and receives advisory or supervisory setpoint adjustments of key-operating parameters.

. It allows engineers and planners to study the profitability of different modes of plant operation.

Once the reconciliation phase is complete, in order to produce coherent data detailed reporting on plant performance is done. These reports are tailored to the requirements of the enduser eg. plant operator, planning department, accounting department etc.

. It optimizes plant operations

APMS also calculates the set points of key operating parameters required for optimal operation.

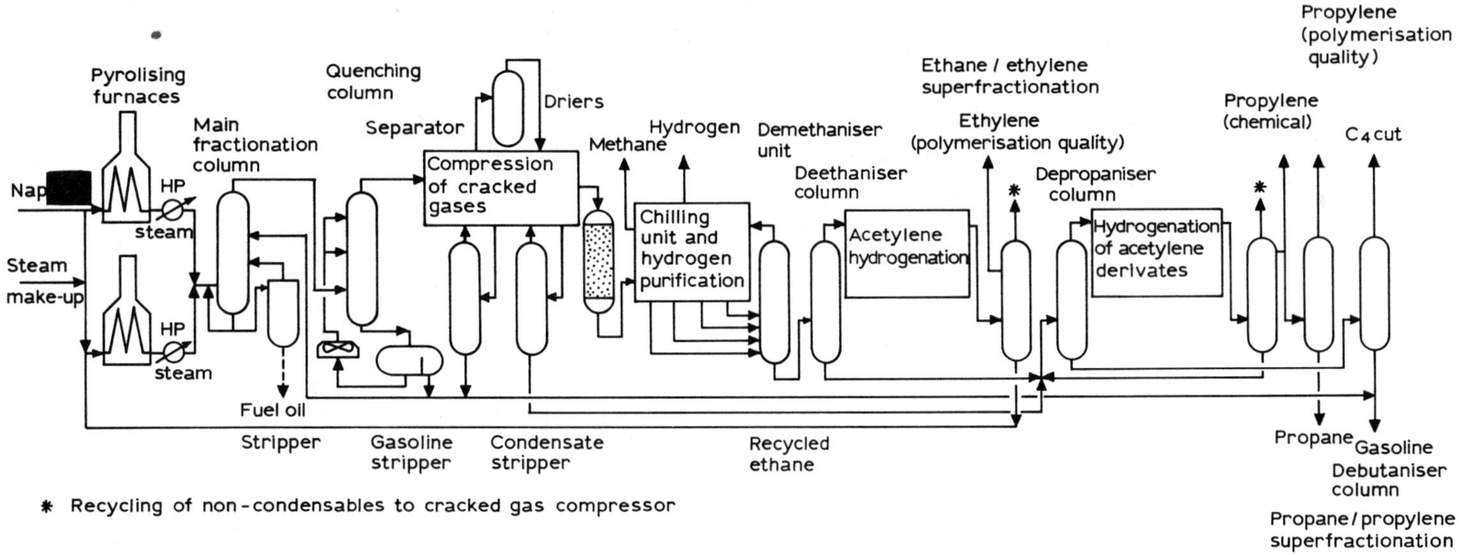

Figure 1 – Typical Ethylene processing scheme

Table 1. - Mass balance and total energy input

Feed	Ethane		Propane		Naphta		Gasoil	
<u>Mass balance</u>								
- Etylene	100	80,0	100	45,0	100	36	100	26
- Propylene	1,7	1,4	32,2	14,5	50	18	52	13
- Butadiene	-	-	6,1	2,8	12	4	16	4
- BTX	-	-	7,4	3,3	24	9	43	11
Sub total A	101,7	81,4	145,7	65,6	186	67	211	54
- C_4 fraction (without butadiene)	6,0	4,8	4,3	2,0	12	4	17	4
- Gasoline	0,3	0,2	9,4	4,2	18	6	38	10
Sub total B	6,3	5,0	13,7	6,2	30	10	55	14
- Fuel gas	17,0	13,6	62,6	28,2	55	20	45	12
- Fuel oil	-	-	-	-	9	3	78	20
Sub total C	17,0	13,6	62,6	28,2	64	23	123	32
SUM TOTAL (Feed)	125,0	100,0	222,0	100,0	280	100	389	100
TOTAL ENERGY								
kWh/t of ethylene	4000		6300		7000		7500	
kWh/t of feedstock	3200		2835		2520		1950	

Table 2. Ethylene cost breakdown in FF per ton

Raw materials cost (naphtha)	3640
By product credit	2766
Net raw material cost	874
Energy cost	588
Total variable cost	1462
Capital costs and benefit	900-1000
Selling price	2360-2460

The optimization is done according to one of a series of objectives contained in the APMS software such as (but not limited to) :
- maximum profitability on variable costs,
- maximum production at given feedstock supply,
- minimum feedstock consumption for a given production requirement,
- minimum energy consumption,
- etc.

Plant limitations and constraints are incorporated in the optimization in order to ensure that optimum settings are achievable.

The standard APMS functions consist of the following :

in on-line mode :

. Reconciliation of measured data,

. Optimization of current operations (either in closed-loop or in open-loop mode).

In off-line mode :

. Simulation/Optimization of plant operations

. Back-up of on-line modes

. Plant operations scheduling

The on line modes may be discarded for those plants not equipped with Digital Control System and APMS functions perform fully off-line providing advisory information to the plant operating personnel and key information to other departments.

3. TISFLO AS FRAMEWORK FOR APMS

For many years, optimization of plant operation has been done with off-line programs using linear programming techniques; besides the obvious limitation of having to represent essentially non-linear behaving plant sections by linearisations over a wide range, such programs only represented the plant in a very global way.

Some current design flowsheeting programs using sequential flowsheeting techniques, are equipped with optimization facilities using the complex or similar methods. These programs are neither suited to on-line nor off-line optimization due to the long calculation times involved and the non-flexibility of such packages for changes in free and fixed parameters. Also data reconciliation and the reflection of plant limitations are difficult to handle. Sometimes the optimization problem is extremely simplified by performing local optimizations.

For the APMS packages, the equation based flowsheeting package TISFLO (Technical Information System) is used.

TISFLO development was started in 1968 within DSM as a simultaneous modular flowsheeting package (an equation based technique was used), however, the non linear models were treated and solved outside the equation system.

In 1979, the TISFLO package was transformed to a fully equation based package enabling complete optimization calculations.

The elementary building blocks of a TISFLO based APMS application for optimization are located in the TISFLO framework :

Physical properties routines

Plant section models

Plant data

Optimization objective function(s)

In essence, the application model for other APMS functions such as the reconciliation package is of the same form. The particular functions are selected setting appropriate flags.

The different plant sections, such as the reactors, compressor stages, columns, cooling systems, etc. are modelled in the form of non linear equations and represented by Fortran 77 subroutines. The same method is used for the set of possible optimization objectives.

Normally, the non linear equations use physical property routines in order to estimate, for example enthalpy and K values.

Use can be made of any reliable physical properties package such as PYROTEC's PHYSCO package which contains proven methods used for petrochemical and refinery process calculations.

The total set of program routines are then integrated into the TISFLO framework to form the APMS application package, i.e. the APMS/O olefin plant optimizer function, or the APMS/E optimizer function for energy household.

TISFLO contains three types of flowsheet-based calculations :

. Simulation

Calculation of the steady state mass and heat balance

. Optimization

Calculation of the steady state mass and heat balance with the addition of an objective function and a set of plant limitations and imposed constraints.

. Reconciliation

Calculation of a set of correction terms applied to a corresponding set of measured data such that minimization of a least square criterion is obtained. The calculation results in a consistent steady state heat and material balance for the set of measured plant data making use of the overall redundancy of measured data and process model predictions.

In off-line applications, the reconciliation mode is also used to tune model parameters to actual plant conditions.

In simulation mode, the total set of non linear equations is constructed by TISFLO from the plant topology (which is at the database level), the set of non linear plant models and the conditions (compositions, temp. pressure) of the known streams (i.e. feed streams, required product specifications, etc). The non linear equations are then linearized and solved using a proprietary modified Gaussian elimination technique. The matrix of the set of linear equations, which is large and sparse, is built up in a compact way, using linked list structures; pivoting is performed in order to ensure calculation accuracy and to avoid overfilling of the matrix. Convergence is reached when the non linear equations satisfy their tolerance criterion.

In optimization mode, the set of non linear equations is extended by equations derived from the active constraints and the objective function and use is made of Lagrange multipliers with proper selected damping terms.

The reconciliation of measured plant data (temperatures, pressures, flowrates, compositions) is an essential part of the APMS functions. Without reconciliation, the results of the optimization function of APMS cannot be judged against the current operations.

In the daily running of process plants operating conditions may vary from time to time. Feedstock availability and type, production requirements, or temporary bottlenecks in the olefin plant itself or in downstream units may necessitate that optimization to a specific objective is done under imposed constraints that are specific to that moment or to the immediate future. The APMS optimization function offers the flexibility of building in a predefined set of possible optimization objectives and to add additional ones at a later stage. Plant limitations (such as equipment constraints) and imposed constraints (such as a production requirements or product purity) are defined in the database at input level and can be specified at any stage.

This feature of the APMS optimization function offers the flexibility required to cope with any particular optimization problem.

Due to equation based nature of the TISFLO structure inside the APMS functions, considerable flexibility exists for changing free, fixed and constrained variables.

The APMS concept allows a number of predefined combinations of fixed, free and constrained variables for each predefined optimization objectives to be selected via a menu driven input structure. Furthermore, alterations and extensions of these can be made at any time, without the necessity of making changes at the TISFLO program level.

4. CRACKING FURNACE MODELING

The cracking section of the plant is by far the most important one with regard to optimum operation. The downstream section has to process the slate of products as determined by the cracker settings and the feedstock type.

Furthermore, with regard to energy consumption, the cracking furnaces can be considered as the plants "workhorses". The modelling of the cracking furnaces therefore has to be done accurately in terms of the predicted effluent slate, fuel consumption, fouling rate and of limiting variables such as outside tubewall skin temperatures.

PYROTEC uses its well known SPYRO program linked with the FIREBOX and CONVEC programs as the standard tool for model data generation. For simulation of the cracker effluent slate other reliable tools may be used such as a client in house yield prediction program.

Regarding the application of SPYRO as a model data generator two approaches or a combination thereof can be followed.

. SPYRO as an on-line model data generator. Whenever a feedstock spec is changed, an automized procedure performs the on line updating of model parameters. A fast executing version of SPYRO is used for this purpose. This requires a user license for SPYRO usage.

. SPYRO as an off line model data generator. During the construction of the APMS package, a model is built using SPYRO as a model data generator for a predefined range of feedstocks and operating conditions.

If required use can be made of other, i.e. in house yield prediction programs to generate both matrices.

In this case, missing parameters if any (e.g. fouling rate) may be added by using SPYRO based predictions.

The cracking furnace model contains four elements :

The radiant coil model consists of non linear equations expressing component yields, coking rate, radiant absorbed duty and coil pressuredrop as a function of hydrocarbon flowrate, steam dilution, coil outlet pressure, cracking severity and fouling status.
Whenever the feedstock specification changes, the on line SPYRO rigorous program automatically generates a new set of equations. In case the SPYRO program is not available on line a different approach is followed :
the so-called proprietary Mixed Component Approach (MCA). The MCA is based upon information derived from SPYRO off-line simulations of a

series of references feedstocks within the range to be covered by the mode, and upon information derived from a series of SPYRO off line simulations in which the individual feedstock components are varied one by one. In this way it is possible to model, for instance the cracking of light to heavy naphtha, including co-cracking with ethane, C_4 fraction, etc, using SPYRO as an off-line data generator.

Firebox Model

The firebox model is the set of equations expressing the radiant efficiency, flue gas composition and temperature as a function of fuel composition, excess air, ambient air temperature and of parameters (e.g. radiant absorbed duty) calculated by the radiant coil model.

The firebox model equations are derived from a series of PYROTEC's proprietary rigorous FIREBOX + SPYRO simulation runs, covering the required range of conditions.

Convection section model

The convections section model constitutes a set of non linear equations expressing duty pick-up of each convection bank as a function of operating conditions (throughputs, temperatures) and of parameters calculated by the firebox model (fluegas temperature, quantity and compositions). The equations are derived from PYROTEC's proprietary CONVEC rigorous convection section simulation program.

Transferline exchanger model

The transferline exchanger simulation model in the APMS software uses a modified version of PYROTEC's rigorous transferline exchanger simulation program (TES).

The model calculates the HP steam production and process outlet temperature and remaining TLE On stream time,as a function of parameters derived from the radiant coil model and of HP steam pressure level.

Other Plant Sections

The plant is divided in main sections, and each is represented by sets of (non) linear equations representing the equipment and by balance equations and stream topology inside the TISFLO frame. The (non) linear equations are automatically generated by rigourous simulation programs such as PRISMA (PYROTEC) or PROCAS (TECHNIP) and are then tested against plant data.

5. DATA HANDLING IN LARGE SYSTEMS

All these simulation or optimization calculations need and produce a large data flow. For the data processing part, the main problems to be solved to obtain a powerful system are the following :

- data acquisition in real time from the plant
- storage of a large set of data
- access to the stored data through queries and their presentation
- dispatching specific production data to several functions systems for use by particular functions.

A few years ago, this kind of requirements were not easy to meet and needed a lot of expensive specific software developements. Nowadays, with the progress of hardware and software technologies, it is possible to use standard systems which are not specifically designed for one application but which are commonly used. They decrease the cost and increase the reliability of the entire system.

Data acquisition

The data acquisition is realized by distributed control system (DCS) with standard links to computers through a high speed communication interface unit of the DCS. The other functions of the DCS are to handle the control loops to detect the alarms, to take the safety actions and to allow the supervisory control of the plant from the control room by graphic displays.

Data storage

The best way to manage and manipulate large amounts of complex interelated data with maximum flexibility is to use a relational data base management system RDBMS. These software packages were developing mainly for business applications and are now powerful, even for realtime applications.

The RDBMS consists of :

- a complete relational data base manager and information control system
- a fourth generation language (4GL) to replace conventional programming
- sophisticated display programming tools to create forms, reports, queries, without coding
- an integrated applications development environment which increases programmer's productivity.

Data access

It is necessary to access quickly to limited set of data inside the large data base ; the RDBMS provides performances for queries and displayes.

The user interface through CRT and key board is completely interactive and features a cursor and menu approach so that once users learn how to use a few keys, they can immediately be operational with the application.

The RDBMS affords to end users the ability to prototype and to maintain large portions of the interactive application by themselves without requiring a programming language knowledge.

Data transmission

Various people which are not directly attached to the plant operations may be concerned by data requested or produced by the APMS. For exemple, the head quarters of the company fixes a monthly production schedule and wants to know the daily production.

Therefore, the APMS is connected to the existing office system, linking together the main production centers and the head quaters. Actually, there is a session a day for the exchanges but in the future, the distributed data bases on heterogeneous systems will allow opened access to the plant data from any location of the network database system.

Hardware and software choices

The hardware and software choices are made after a large consultation with the main following criteria :

- the computer shall be powerful for the modeling calculation (benchmarks were done with TISFLO)
- the computer shall offer the most standardized connection to the existing DCS.
- the RDBMS shall be standard and compatible with the selected computer
- the RDBMS shall be very efficient for data sorting and incorporate production tools.

The computer selected is a HEWLETT PACKARD 9000/840 which is a very powerful Reduced Instruction Set Computer (RISC).

The RDBMS selected is INGRES which is used for a wide variety of production applications through the world with more than 2500 installations overall.

6. WHAT IS TO BE EXPECTED FROM APMS ?

Let us recall the main technical reasons for implementing an APMS on an ethylene unit :

A. Multiple products with some possibilities to adjust relative production rates, but with many interactions and constraints.
B. Fluctuating market demand, storage capacity must be used for balancing.
C. Changing feed quality, having an impact on capacity and relative product output.
D. Parallel operation of multiple (8-12) reactor units (cracking furnaces).
E. Big impact of energy cost on total production cost.
F. Large number of setpoints to be defined and monitored (120-150 for the furnaces only).

It is now possible to relate each function to these characteristic operating constraints and describe briefly their advantage with respect to traditional control systems.

Scheduling - Typical time span : 1 month

The value of the scheduling function is to forecast the operating conditions over longer time periods based on anticipated feed availability and product demand. The model of the unit allows a close look at the furnace operating conditions. Traditional systems apply linear programming and are not related back to the unit operating conditions. Anotehr advantage of APMS is the possibility to update the forecasts when new information is available, by recalling the latest schedule from the data base.

Various scenarios can be tested yielding material, energy balances and economic criteria.

Balances, Status Report

Typical time span : 24 Hours

From the order of magnitude 1000 data points to be collected, this function extracts synthetic reports such as coherent material and energy balances and characteristic parameters related to the equipment performance. This is possible by applying the plant model to the measured data and calculating the smoothed parameters over 8 and 24 hours. This gives a historic base in order to evaluate the recent plant performance and follow larger term trends on unbiased data basis.

Traditionnaly some important data were collected daily and put on graphs, such as furnace tube metal temperature or quench boiler outlet temperature (characteristic of furnace coking/expected runlenght).

With the APMS the operator has various parameters at hand and gains an enormous insight in the actual unbiased performance. Also status reports to higher management levels are easy to edit.

Optimization/targets : Time span 4 hours

From present value operating data, such as feed quality, feed and/or product rates, ambient temperature, furnace conditions, the plant model is run in an optimizer mode, issuing setpoints (targets) for the main controllers. These set points are in line with certain constraints such as ethylene/propylene ratio, maximum furnace outlet temperature, and yield an optimum of some objective function. This latter is generally the gross margin.

Here again the size of the problem prevents any approach by traditional systems. Also a very important aspect is the capability to foresee trends on the economic function and/or on the technical constraints when certain specific parameters are changed. Some of these, i.e. dilution steam ratio and gas compressor suction pressure are interrelated and their effect on the performance is particularly implicite and quite impossible to predict. Many other parameters, such as reflux ratios, are similarly very indirectly related to the energetic performance of the overall plant, and can be better monitored through the APMS.

Advanced Control : Time span 10 minutes
This is the domain of the local optimizers which keep subsections or main equipment items at best performance considering only local conditions. Such systems are however a large improvement over the single loop control concept. They can control the ultimate parameter such as a heat input, rather than an inferred parameter, such as a temperature of a heating stream.

To predict the expected gain from an APMS system we use one of severall possible approaches.

First of all actual plant data from test runs and from current operations are collected. Their evaluation is done according one of the following methods :

- rigorous simulation of the plant is taken as a reference,
- "the best operator" performance is taken as a reference,
- typical performance values for similar plants are taken as a reference,
- a statistical analysis of a large sample of operating performances supplies "the best expected" values.

We think that the second and the fourth method are most often applicable, and can be used to generate a basis for guaranteed performance after implementation of the APMS.

On a 1000 t/d ethylene unit it was possible to save 15 tons of fuel oil equivalent and produce 2 tons of ethylene more per day of operation by the implementation of an Advanced Control System only. The total energy consumption of this plant was 750 tons of fuel oil per day. This saving might appear marginal, however the pay out time including all plant hardware related to the project was less than two years. Considering in addition what it means to avoid only one 12 hour production outage a year (there are about 6 per year in a traditional plant) due to the better performance monitoring, it is easy to conclude on the value of APMS.

As a first approach it is recommended to install an Advanced Control System and an offline model to perform the scheduling funtion. The complete APMS can then be implemented as a second phase project, including product movement storage and utility systems.

MICROPROCESSOR SYSTEM AND DIGITAL REGULATION LOOPS FOR INCREASING COWPERS ENERGY SAVINGS

A. SCIARRETTA
Process Control of Pig Iron Area
at Italsider Taranto Steel Works

1. PREFACE

Italsider is the most important iron and steel works in Italy. It is a company operating in the context of the State ownership and it is under Finsider, a Iri holding for the steel industry.

Italsider concentrates its activity in the field of flat hot, and cold rolled products, of coated rolled products (zinc coated, tin plates and chromium plates), of medium and large diameter welded pipes.

This company consists of two integrated iron and steel plants, set by the sea (Taranto and Bagnoli), and four more minor productive units.

The Italsider Steel Works has, in Taranto, a productive capacity of 11.5 millions tons of steel per year. It produces flat hot and cold rolled products (wide and narrow strips, heavy, medium and thin plates) and it is specialized in producing large diameter longitudinal and spiral welded pipes.

The whole plant consists of:

- 5 large diameter blast furnaces
- 2 steel making plants with 6 LD 350 tons converters
- 5 continuous casting plants
- 1 plate mill
- 2 hot strip mills
- 1 cold rolling mill for wide strips
- 2 longitudinal welded pipe mills
- 2 spiral welded pipe mills

2. INTRODUCTION

For Italsider Taranto iron and steel plant, Blast Furnace No. 5 (picture I) means a very important step in its technological development. Design, construction and use of this blast furnace are the final result of research and technical improvements due to a substantial effort of the company.

Picture I - Taranto Steel Works: Blast Furnace No. 5

A blast furnace is big not only for its volume but fundamentally for all the several used technologies.

Blast furnace No. 5 has been started up on November 20, 1974 and it completed its first campaign on September 11, 1981, producing about 20 millions tons of pig iron. The re-lining works lasted 276 days and on 24 June, 1982 the plant has been started up for the second campaign. The present production is 9,200 pig iron tons per day with a consumption, in the hot stoves, of 480 MCAL/TON of pig iron. (It was 560 MCAL/TON of pig iron during the first campaign). (Figure I).

Through the re-lining important innovations have been introduced in order to improve:

- the duration of the blast furnace campaign;
- the reduction of the shut down period for maintenance works;
- the ecology of the work environment;
- the energy consumption.

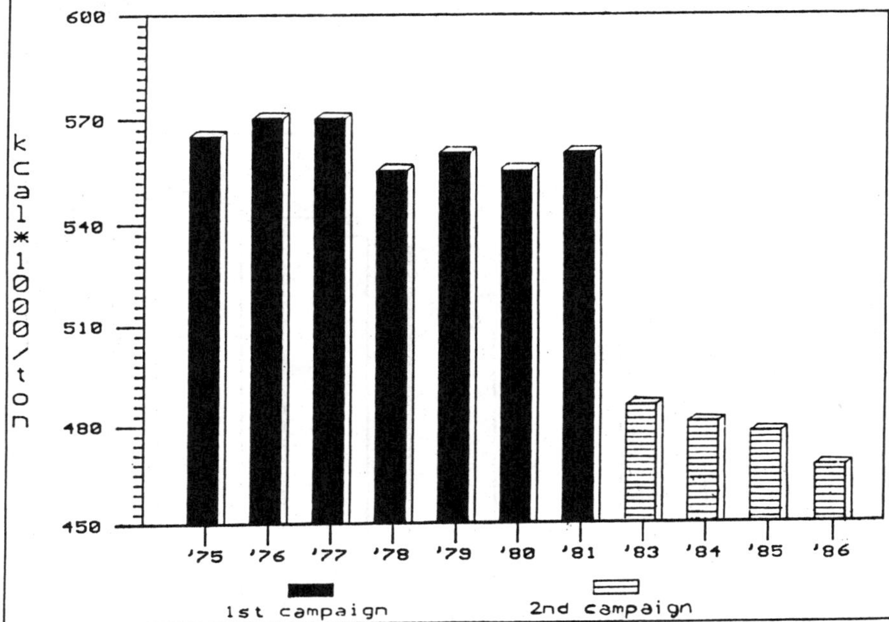

Fig. I - Energy consumption in the hot stoves

The main innovations concern:

- Paul Würth throath equipment;
- SIC (silicon carbide) refractory in the most thermically stressed areas (bush, belly, half stack);
- installation of the turbine to recuperate energy from top pressure;
- improvement of the plant and the hot stoves management.

At the moment Italsider is defining and planning a scheme which will make available a system of automation OOCS (Overall Operation Control System); the final aim is the operation of the blast furnace through expert systems (artificial intelligence).

One of the most important innovations, as far as energy savings are concerned, is the hot stoves plant. Its efficiency, which was 68-70% during the first blast furnace's campaign, has improved thanks to the following changes:

- installation of new ceramic burners;
- combustion control with the enrichment of the blast furnace gas for each hot stove;
- recovery of the waste gases heat;
- automatic operation of the plant through microprocessors and process computer.

All these changes have made possible an increase of 5-7% in the performance. (FigureII).

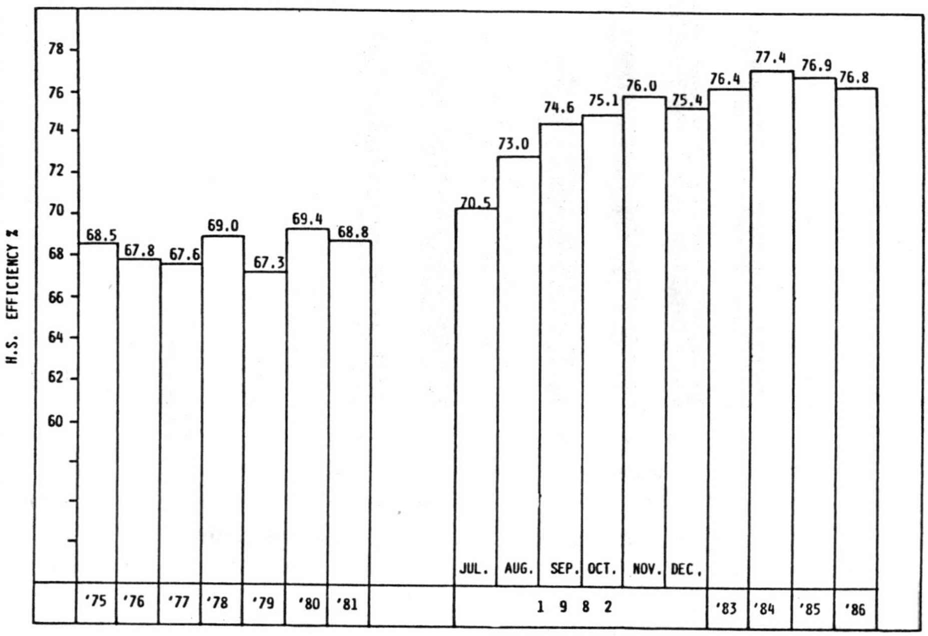

Fig. II - Taranto 5 B. F. hot stoves efficiency

3. <u>DESCRIPTION OF THE HOT STOVE PLANT</u>

The blast heating plant of this blast furnace includes four hot stoves. Each of them is made up of two main parts: a combustion chamber and a checker chamber. The fuel (a mixture composed by lean gas,that is to say blast furnace gas,and rich gas, coke oven gas or natural gas) and the combustion air are introduced into the combustion chamber (figure III).

Both the fuel and the combustion air are pre-heated using the heat of the fumes escaping out of the hot stoves (figure IV). In the checkers chamber the heat energy, produced during the combustion (called gas phase or re-heat) accumulates, thanks to the passage of the hot gas through a column of hollow bricks, which form the checkers. The energy accumulated is then transferred to the cold blast, which passes through the checkers chamber to supply the hot blast to the blast furnace, (called blast phase or blowing).

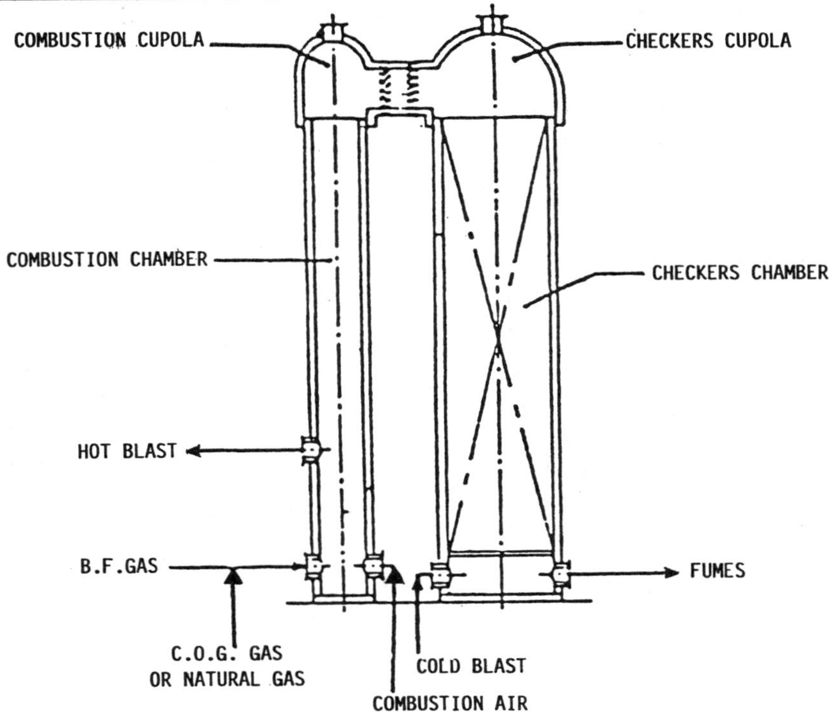

Fig. III - Hot stoves semplified scheme

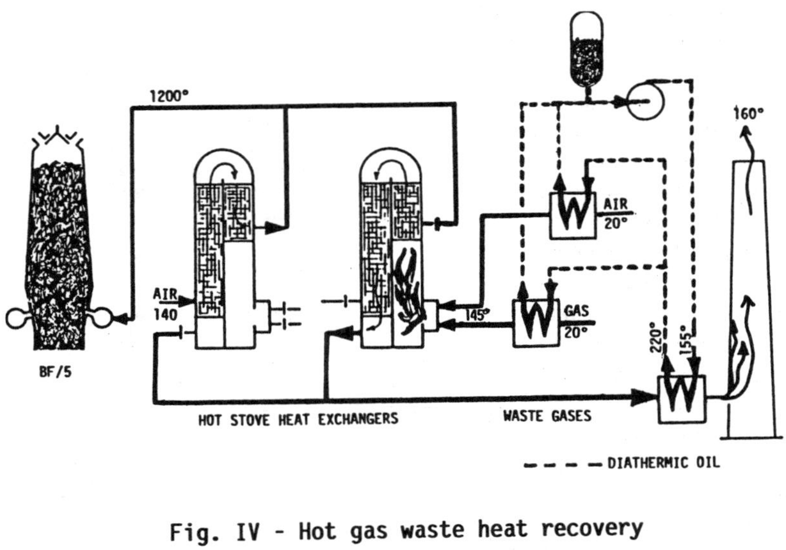

Fig. IV - Hot gas waste heat recovery

4. TECHNICAL FEATURES

Before realizing the demonstrative project, as far as the control of the combustion, the phases and the reversing are concerned, both the traditional technology (analog adjustment and assembled flows) and the manual mode, controlled by the operator, were employed to operate the hot stove plant. This new system can manage the plant by the automatic mode; it employes either the microprocessors to control the valves directly and check the adjustment loops, or the process computer, connected to the microprocessors via teleprocessings to supervise and optimize the process (see the illustrated lay-out, fig. V).

It is clear why we have chosen a running with microprocessors instead of one where the process computer was utilized to direct the adjustment and start the sequences. The advantages which could derive from a process computer are similar to those ones offered by a centralized adjustment system which can use adjustment algorithms even more sophisticated than PID and which can have interconnection among the adjustment loops, implemented with a software.

The disadvantages which came out derived from the following elements:

- a system committed to the working of an only component (computer), therefore needing some redundant systems;
- presence on the market of more modern equipments,

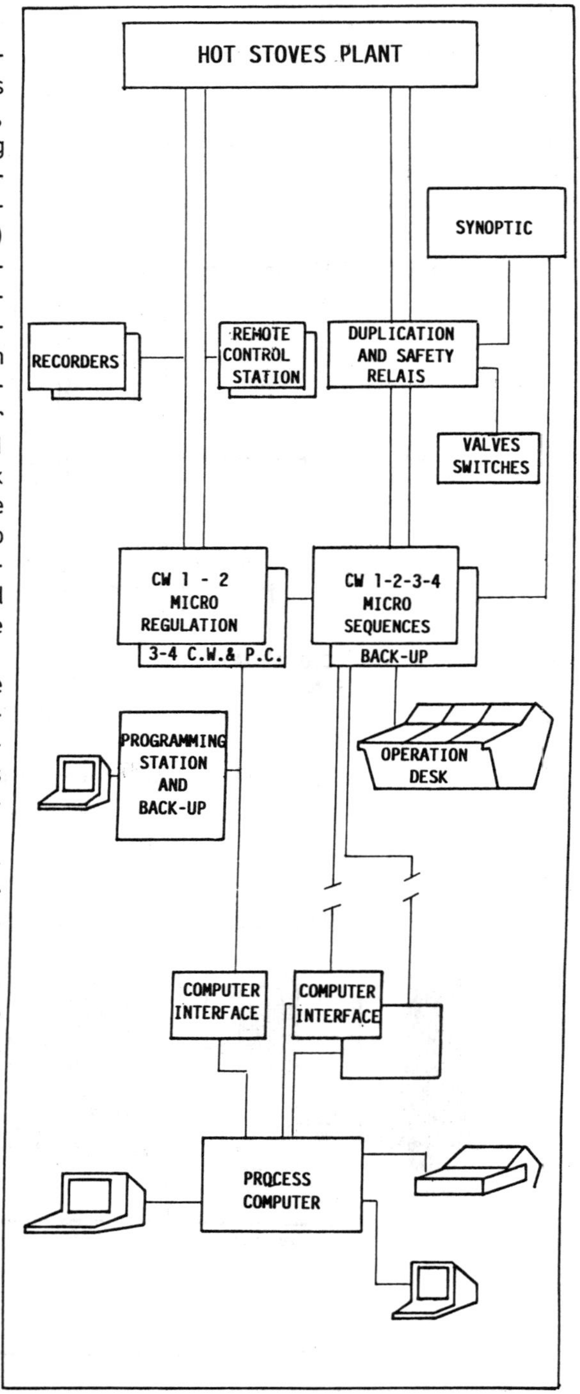

Fig. V - Project lay-out

which are more suitable to the man/plant needs, and can be more easily interfaced with the plant;

- consolidation of the digital technic in the production of automatic regulators with high operational speed and with costs close to those of the traditional analogue regulators.

Our system gives the opportunity to distribute the automation functions on two levels:

1. supervision and models committed to the process computer
2. automation and regulation committed to the microprocessors and a greater safety in the use of the some main functions, when some parts of the plant are not available, too.

The expected energy savings are connected to the effects, taken as a whole, of an improved process control, achievable through the microprocessors and the process computer. But the main advantage for the energy savings are connected with the new strategy used in the hot-stove re-heating (gas phase), which has been accomplished thanks to this new system. Previously, in fact, after having reached the dome temperature (in about 7-10 minutes, its set being usually fixed at 1,400°C), in the remaining time of re-heating phase (40-45 minutes, being the total time of the gas phase 50-52 minutes), the necessary decreasing of the flame temperature was obtained by mantaining the contributions of the fuel gas unchanged (the B.F. gas and the C.O.G. flows) and increasing the combustion air flow (adjustment with air-excess, figure VI).

On the other hand the adjustment estimated and accomplished with the new system consists in decreasing the flame temperature, after having reached the set fixed for the dome temperature; the combustion air flow is maintained unchanged, but the C.O.G. flow is reduced and the B.F. gas flow is increased. In this phase, too, the fuel-air ratio is kept close to the stechiometric ratio. (New adjustment with reduction of rich gas, fig.VII).

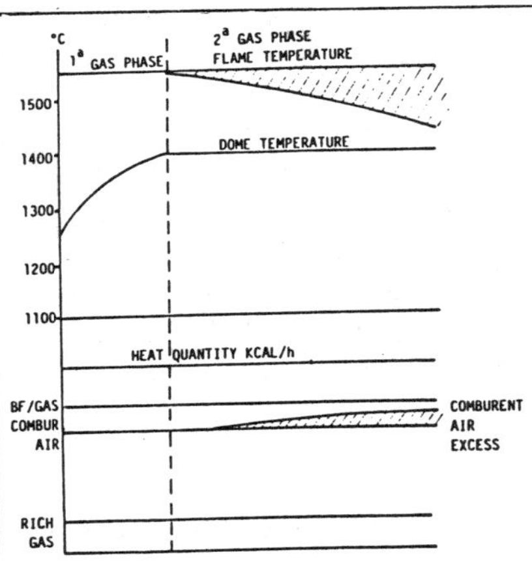

FIG. VI - Heating phase adjustment with air excess

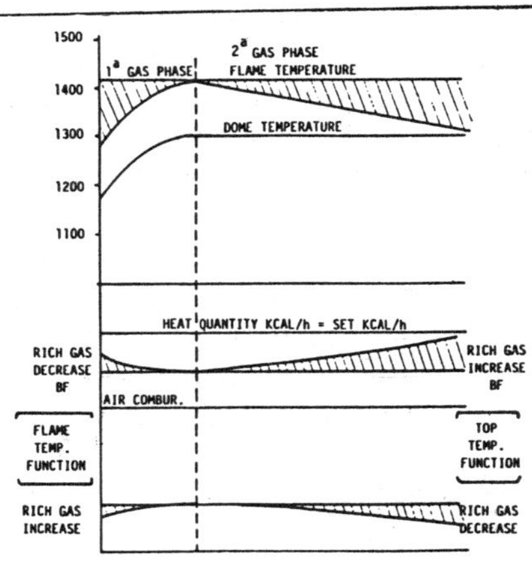

FIG. VII - Heating phase adjustment with reduction of rich gas

5. USED EQUIPMENTS

The equipments used for the operation are:

- micro regulations
- micro sequences or inversions
- process computer

Micro regulations: when the micro regulations which control the adjustments have received the set values, they can set indipendently the values for the re-heating and blowing phases.

These micro regulations are two, one for the adjustment of hot stoves 1 and 2, the other for the adjustments of hot stoves 3 and 4 and that one related to the common part of the plant.

The adjustments affecting the energy savings and taking part in the re-heating phase are:

- adjustment of blast furnace gas;
- adjustment of rich gas (coke oven gas or natural gas);
- adjustment of the dome temperature
- adjustment of the combustion air

Adjustment of the blast furnace gas: the micro regulation calculates the set value required for the blast furnace gas flow according to the adjustment sets (thermic capacity, percentage of rich gas, heat of combustion of the blast furnace gas and of the rich gas). These values are set by the operator through the display or automatically supplied by the supervising computer by means of mathematical models.

Ajustment of the rich gas (coke oven gas or natural gas): the ratio regulator adjusts the flow of mix gas rich gas according to the set value of percentage of rich gas.

This value, set up at the beginning of the re-heating phase, is modified by the mathematical models before it reaches the dome temperature set, according to the ascent time and the flame temperature needed; it is then modified by the micro regulations, through the dome temperature regulator, in order to keep this temperature at set value.

Adjustment of the dome temperature: the adjustment of the dome temperature interacts with the set value of the percentage of enrichment (it increases or decreases); therefore it interacts both with the adjustments of the blast furnace gas flow and of the rich gas flow.

Adjustment of the air to gas ratio: the ratio set required for the air/mix gas is calculated according to the stechiometric sets of the air/blast furnace gas and air/rich gas and according to an optimization coefficient (set by the operator).

Micro sequences or inversions: the micro wich control the adjustments of the inversions are two: one of them acts on the plant, the other is for the back-up.

When they receive the signal to start the sequence, they can autonomously bring the hot stove to the required state (gas, wind or closed), controls the valves interested in the process and check the time required for the manoeuvre. They visualize the valves and the hot stove's state on the synoptic, intervene for a failure or an alarm and put the plant into safety.

Process computer: the supervising process computer controls the re-heating and the blowing phases, using proper mathematical models.

Through these it is possible to transmit the set values to the regulations micro and the start for the sequences to the sequences or inversions micro.

The most significant targets obtained with the use of the mathematical models and the supervision of the operation are:

- the optimization of the re-heating phase by checking the amount of heat to be supplied to the hot stove;
- the adjustment of the enrichment of percentage, before it reaches the dome temperature, getting an optimal ascent curve of the flame temperature;
- the managing of the duration of the re-heating phase;
- the managing of the transition phases due to:
  - variation of the required heat (set value of the temperature, hot wind, wind flow...);
  - heat shortage due to some anomaly.

6. RESULTS OF THE PROJECT

A comparison, between the present consumptions and the pre-existing ones, must take into account other energy saving undertakings that have been achieved in addition to this project, for example the installation of new ceramic burners or a plant to recover the fumes' heat.

The quantity of the saved energy has been obtained comparing the present arrangment (adjustment and decrease of rich gas) with the previous one (adjustment with air excess): for this comparision the so-called "indirect method" has been used. It consists in calculating not the difference of the consumption, but the difference of the losses in the two mentioned arrangements.

It is important to mention the installation on the hot stove's discharge of a meter to measure how much O_2 there is in the fumes (the oximeter EC18G, an absorption instrument made by Hermann Moritz).

The measurements, registered minute by minute, have immediately shown how, during the second re-heating phase (after the dome temperature had been reached) the new adjustment, with a decrease for the rich gas, gave a 0.25 % O_2 in the fumes; on the contrary with the previous adjustment with air surplus the O_2 obtained value was about 1%. With regard to the two fuels' composition and the blast furnace gas/coke oven gas ratio, these two values coincided with the air surplus, getting respectively 2-3% (therefore optimal) and 10-11% air surplus.

Besides, in the first case (adjustment with decreasing of rich gas) the fumes temperature has been about 350 °C, while, in the second one (adjustment with air excess) it has been 5°C more.

The energy saving which can be achieved is therefore assessable just adding these two following effects (they are related to the gas phase's period, after the dome set has been reached):

-smaller flow of the fumes;

-lower temperature of the fumes going out the hot stoves.

In fact, as far as the wind phase is concerned, the new system using the microprocessors makes no difference about the energy aspect. The energy saving which can be obtained is therefore 1,160 Tep/year.

The system is very efficient, reliable, suitable to ensure an exact adjustment and minimize the energy losses.

7. ECONOMIC ASPECTS

The capital cost of the project, started in 1981 and finished in 1984, was 1,352 millions lit.: 1,091 millions lit. have been paid for the equipments and 261 millions lit. for measurements campaign. The pay-back is achieved within 4 - 7 years.

SESSION V : ENERGY MANAGEMENT OF UTILITIES

New technics for the management of utilities in industrial plants

Application of the SECI-MANAGER software to energy systems optimization and on-line industrial processes

Energy savings and economic consequences resulting from the installation of a cogeneration unit (electricity - steam) at the Corinth refinery (Motor Oil Hellas)

Software systems to optimize combined heat and power plant

NEW TECHNICS FOR THE MANAGEMENT OF UTILITIES IN INDUSTRIAL PLANTS

G.B. ZORZOLI
Board of Directors
ENEL - Rome

1. INTRODUCTION

Today any kind of energy management should aim at two targets : i) a high flexibility with respect to fuels ; ii) a low environmental impact. These targets are consistent with current trends both in the energy sources market and in the public opinion top priorities. The two targets could be contradictory in as much the former can be achieved by means of a high flexibility in the fuel to be burnt, which in turn may lead to exploit highly polluting sources. The contradiction can be overcome either by improving the overall efficiency of the energy system or by adopting advanced technologies capable of dealing with polluting fuels. The combined use of both strategies can be incorporated in the same technical solution. A strong support also comes from adopting advanced solutions in the field of control systems.

A comprehensive survey of new technics developed for industrial application, is a hard task, due to the wide range of technical specifications present in the industrial sector. As a matter of fact there are industrial plants where energy demand is mainly due to space heating, whereas elsewhere process heat or electric power can dominate supply problems. Therefore in what follows a reasonable selection among any new technics has been chosen.

2. TECHNOLOGIES FOR UNCONVENTIONAL SOURCES

Fluidized-bed combustion boilers are at present the most promising solution to burn a large variety of fuels, even if highly polluting. Moreover the fluidized-bed concept is inherently appropriate to the size range that industrial applications can require.

Two solutions are now available : the Bubbling Fluidized Bed (BFB) and Circulating Fluidized Bed (CFB). Both BFB and CFB offer distinct advantages over traditional pulverized or stoker coal boilers. The technical viability of both technologies has been proven by full scale operating units, and general acceptance of the fluidized bed concept is growing. Both designs allow the use of low grade, hard-to-burn fuels without the need for expensive post-combustion scrubbing equipment. However, as shown in Table I, there are differences between BFB and CFB in cost, operation and application. Generally speaking CFB has proven itself superior to BFB in many ways. In addition to CFB's better performance in combustion efficiency, sulphur capture is significantly greater with less limestone required. Its simpler fuel feed system and lack of in-bed tubes mean lower maintenance and operating costs. A CFB's biggest disadvantage is its higher cost for smaller units and its relatively large vertical height requirement. A BFB offers some advantages in certain cases and, therefore, must not be forgotten.

When the prime mover is a gas turbine (see chapter 3), only high quality conventional fuels can be exploited, such as natural gas, gas oil, to some extent low-sulphur fuel oil. In order to overcome this bond, coal gasification has been investigated. Several processes are now being developed : one of the most promising will be described here (2).

At the gasifier, which is a refractory-lined vessel mounted at the top of the radiant syngas cooler, a coal-water slurry is combined with oxygen supplied from an adjacent air-separation plant. Mixed in a specially designed burner, partial oxidation reactions take place to produce a medium Btu syngas consisting mainly of CO, H_2, CO_2 and steam. (Clean syngas composition is shown in Table II). Most of the sulphur is converted to H_2S, with a smaller amount of COS formed. The hot syngas and slag from the gasifier reactor discharge into the radiant cooler, which generates high pressure saturated steam. The slag drops into a water pool at the bottom of the radiant cooler and is removed through a lockhopper system. The syngas proceeds into a convection cooler where additional high pressure saturated steam is generated.

From the convection cooler, the gas enters a carbon scrubber, where essentially all of the fine particulates are removed and recycled. After further cooling, the syngas proceeds to a unit, which removes 97 % of the sulphur. The clean, dry syngas (at approximately 2600 kcal/m^3) is then contacted with water in a saturator prior to firing in the gas turbine. The water provides moisture addition that is needed to control nitrogen oxide (NOx) emissions from the gas turbine. Alternatively, NOx emissions can be controlled equally well by the direct injection of steam at the combustion turbine.

Since the gas scrubbing is at low temperature, energy losses cannot be avoided. Therefore studies aimed at utilizing hot scrubbing technics are now in progress.

In the near future an abundant "unconventional" energy source, i.e. low temperature heat, could be widely upgraded for industrial applications by means of heat pumps (3). The industrial heat pump, however, has to compete with several alternative technologies, such as modern boilers, regenerators, recuperators, and heat pipes, with regard to profitability, primary energy savings, recovery of waste heat and reduction of environmental impact. The following requirements must be met for the industrial heat pump to prevail against these alternatives :

1. manufacture of industrial heat pumps with lower first cost ;
2. development of heat pumps with output temperatures in the range of 150-300°C, temperatures necessary for many industrial processes ;
3. intensive and detailed analysis of different types of industrial heat pumps for specific branches of industry ;
4. better adaptation of process technologies to heat pump applications.

Several techniques are now being used or are under development for high-temperature industrial heat pumps. For output temperatures above 150°C, only heat transformer, mechanical vapour compressor and steam jet compressor are currently available technologies. The search for suitable refrigerants, refrigerant mixtures and more effective compressors for compression heat pump is continuing. In the case of sorption system, work is now being done on new working fluid mixtures, improved component design and more effective system schematics.

Since cases where heat sources around 50°C are available and industrial end uses range between 70-90°C are not uncommon, even today's high COP heat pumps can have a reasonable number of applications.

Fuel cells can represent a more advanced technology particularly suited to industrial applications (4). The main type of fuel cells, along with their characteristics, are listed in table III. Fuel cells can grant :

1. an electrical efficiency as high as 40-45 % when phosphoric acid technology is adopted, up to 60 % in the case of molten carbonate technology ; such efficiency is independent of unit power ;
2. combined heat and power generation without any decrement in the

TABLE I

BFB/CFB COMPARISON

	BFB	CFB
Steam production	As low as practicable	50-1500 t/hr
Combustion efficiency	Usual standard	2-3 % better
Heat recovery	Inbed tubes	No inbed tubes
Sorbent use	Up to 100 % more	Usual standard
O & M	Usual standard	Lower
Cost	Similar	Similar

TABLE II

CLEAN SYNGAS COMPOSITION

Component	Mol % (Dry Basis)
CO	42.5
H_2	38.2
CO_2	18.6
CH_4	0.3
Ar & N_2	0.4
H_2S & COS	50 PPM

TABLE III

FUEL CELL CHARACTERISTICS

TYPE	Alkaline electrolyte	Phosphoric acid	Molten carbonate	Solid oxide
OPERAT. TEMP.	60-120°C	150-230°C	600-700°C	800-1100°C
ELECTROLYTE	Potassium hydroxide	Phosphoric acid	Alkaline metalcar-bonates	Zirconium oxide
CATALYST	Platinum, Palladium, Nickel	Platinum	Nickel	---
APPLICATION	Space, porta-ble genera-tors, trans-port	Power plants, small genera-tors, CHPG	Power plants, CHPG	Power plants, CHPG
COMMERCIAL UTILIZATION FORCAST	Late eighties	Around 1990	Late nineties	After 2000

electrical efficiency, so as to achieve an overall efficiency up to 90 % ;

3. a very low environmental impact, since fuel cell effluents have a very low content as long as NOx, SOx, particulates and unburnt hydrocarbons are concerned ;
4. a modular concept, which makes it possible to match the wanted power simply by piling-up the appropriate number of fuel cell modules ;
5. a large number of primary energy sources, by simply changing the reforming unit : natural gas, liquid hydrocarbons, coal.

As heat pumps, fuel cells are handicapped by the availability of proven technologies, which can guarantee at reasonable costs the most significant performance required by industrial applications.

3. COMBINED HEAT AND POWER GENERATION (CHPG)

CHPG is a highly flexible process : it can be supplied by several fuels, even local wastes, it can exploit different technologies and thermal cycles, it can operate at different power yields and/or supply heat having different enthalpy levels. Its high overall efficiency helps achieving an acceptable environmental impact. CHPG inherent flexibility also creates a strong push towards technological innovation and its extension to new applications.

Possibly the best known method of providing (electric) power and process heat is the use of back pressure steam turbines. The steam is used first to generate power and then, after exhausting from the turbine, as process steam. This system in its simplest form offers high thermal efficiency, since all the heat not converted to work in the turbine appears as process heat. The only net losses from the system are the boiler stack loss and the parasitic leakages. The overall efficiency can therefore be around 80 %. To maintain this, clearly all the steam must be used and turbine is normally sized to suit the process steam requirement. In this sense, the power is generated as a by-product of the heat requirement. Since process steam is required at pressures above the atmospheric, the turbine expansion ratio is curtailed, and the Rankine cycle efficiency is severely reduced, relative to a condensing cycle. This results in a high ratio of process heat to power, typically 8 or more.

Recovery of the waste heat from reciprocating engines is also possible. Since these engines have thermal efficiencies up to 40 % and the heat rejected appears in the cooling water, in the exhaust gas and in the lubricating oil, it can usually provide a ratio of heat to power of the order of 1 or less.

The gas turbine has a thermal efficiency intermediate between that of the back pressure steam turbine and the reciprocating engine. Since the heat rejected all appears in the exhaust at a useful temperature of 450-550°C, it can be utilised more simply than in a reciprocating engine. As might be expected from these simple thermodynamic considerations, the gas turbine system occupies the middle ground between the steam turbine and the reciprocating engine. This is illustrated in Table IV.

The most innovative solutions are in the field of gas turbine and reciprocating engine based systems. Table V shows some unusual fuels which can be used in a gas turbine generator set developed for CHPG's (5). Diesel engines for CHPG systems, suited to burn low-heat content gas from wood, were also developed (6). The like also stands for burning heavy fuel (7).

The adoption of automotive engines, which was pioneered by the TOTEM system, based on a 127 FIAT car engine, now lists several solutions, which allow the use of CHPG systems in the very low power range (15 to 2000 KW) (8).

TABLE IV

BASIC HEAT/POWER REGIMES

	Reciprocating Engine	Gas Turbine	Steam Turbine
Cycle Effy %	35–40	15–25	8–12
Typical Heat/Power Ratio	1	2.0–4.0	6–9
Overall Thermal Effy %	60–70	55–80	75–80

TABLE V

SOME UNUSUAL FUELS FOR GAS TURBINE GENERATOR SETS

Source	Nominal Heating Value
Waste water treatment plant digester gas	4600 (Kcal/m^3)
Solid waste landfill gas	3500 "
Chemical process off-gas	1100 "
Coke oven gas	3700 "
Blast furnace gas	1200 "
Coal mine vent gas	3700 "
Bio-waste digester gas	4600 "
Bio-waste gasifier gas (air blown)	1500 "
Municipal waste gasifier (air blown)	1500 "
Coal or coke gasifier (air blown)	1600 "
Oil fields and refineries, crudes and distillates	9900-1500 (Kcal/kg)
Well head condensate	12100 (Kcal/kg)

TABLE VI

TURBINE EXHAUST USED AS COMBUSTION AIR IN 14 ATM BOILER

Main Characteristics	Case 1	Case 2	Case 3
Stack Temperature (°C)	150	150	150
Steam Output (Kg/hr)	15,543	44,228	91,930
Additional Fuel (million Kcal/hr)	7.53	21.9	45.9
Exhaust Temperature (°C)	453	428	412
Fuel Input (million Kcal/hr)	3.33	10.04	20.67
Electrical Output (KW)	800	2706	7209
Air Mass Flow (t/hr)	22.32	63.50	132.09
Net Fuel Rate (Kcal/kWh)	1099	1278	1036

TABLE VII

TURBINE EXHAUST USED TO PRODUCE 1 ATM STEAM WITH NO ADDITIONAL FUEL BURNED

Main Characteristics	Case 1	Case 2	Case 3
Stack Temperature (°C)	150	150	150
Steam Output (Kg/hr)	2914	7649	14,929
Exhaust Temperature (°C)	453	428	412
Fuel Input (million Kcal/hr)	3.33	10.04	20.67
Electrical Output (KW)	800	2706	7209
Air Mass Flow (t/hr)	22.32	63.50	132.09
Net Fuel Rate (Kcal/kWh)	1644	1755	1435

Another innovative stream is related to unconventional thermal cycles. Simplified combined cycles were conceived, so as to be available on a pretty small scale (9). Table VI and Table VII show the main characteristics of similar cycles with additional and no additional fuel burned.

when low enthalpy recovered heat is the energy source, problems to be solved are not only related to thermodynamics difficulties, but also to the possible pollution coming from the technological process (crud, corrosion agents and so on). Both effects lead to larger heat exchange surfaces, therefore to higher costs. The most innovative effort, however, comes from the development of thermal cycles based on fluids different from water vapour, mainly organic fuels (10).

The potential for these solutions is limited by :

1. standardization problems, because solutions must fit in different technological processes, each having specific characteristics ;
2. the above-mentioned large heat exchange surfaces ;
3. a very short pay-back time, because of the limited life expectation due to routine modifications, innovations, replacements in the technological process they fit in.

4. ADVANCED CONTROL SYSTEMS

Microprocessor control systems, mainly for cogeneration packages, are reducing maintenance and operating costs and improve the interface of the CHPG system with the local load and the electric grid. The system can be made as unmanned as possible. In order to achieve this goal a Programmable Logic Controller (PLC) can be installed. The PLC can easily be programmed (and reprogrammed if necessary ; this is one advantage over electro-mechanical relay systems) to initiate events in a required sequence, to check conditions and to route the sequence accordingly.

Of course the availability of updated hardware is not so important as the procurement of an appropriate software, both in the design stage and in the operation phase (11).

In a near future the wider and cheaper availability of expert systems will also help the management of utilities in industrial plants.

Last but not least, the development of new sensors and sensor applications will improve the capability of optimizing the management of such utilities. The potential of these techniques is well illustrated by a tunable laser system, which is measuring in real time the chemical composition inside a combustion chamber, compares it with its optimal composition and consequently acts on the combustion system control. Now applied in steel mills, such a system could have further applications. This example shows how the good management of utilities might be enhanced in the near future.

As a general conclusion, one can forecast that innovations in the field of microelectronics and information technology will probably improve the management of utilities in the industrial plants at least as much as the future achievements in the conventional hardware.

REFERENCES

(1) GAGLIA B.N., HALL A. (1987). Comparison of bubbling and circulating fluidized bed industrial steam generation. International Conference on Fluidized Bed Combustion, Boston.

(2) CLARK W.N., SHORTER V.R. (1986). Cool Water : Mid-term performance assessment. VI Annual EPRI Coal Gasification Contractor's Conference, Palo Alto.

(3) SEVERAL AUTHORS (1986). H.P. Newsletter. Vol. 4, No. 2.

(4) VELLONE R. (1987). R and D and Demonstration program on fuel cells in

Italy. International Seminar on Fuel Cell Technology and Application, The Hague.

(5) BECKMAN J.W. (1984). Cogenerating with unusual fuels in gas turbines. Cogeneration World. Vol. 3, No. 3.

(6) (1986). Low-Btu gas from wood to fuel 20 MW diesel CHP power plant. Cogeneration. September/October.

(7) KIMSTRA K., KEY J. (1987). A highly efficient diesel total energy system burning residual fuel. Cogeneration World. Vol. 6, No. 1.

(8) STAMBLER I. (1985). Reciprocating engines dominate California's small cogeneration plans. Cogeneration. September/October.

(9) JEFFS E. (1983). CIMAC conference highlights design for alternate fuels. Gas Turbine World. July/August.

(10) ANGELINO G., MORONI V. (1973). Perspectives for waste heat recovery by means of organic fluid cycles. Journal of Engineering for Power. April.

(11) COEYATAUX M. (1987). Applications of the SECI MANAGER software to energy systems optimisation and on line industrial processes. This Conference.

SPRINGELL J. (1987). The use of software systems to optimise CHP. This Conference.

APPLICATION OF THE SECI-MANAGER SOFTWARE TO ENERGY SYSTEMS OPTIMIZATION AND ON-LINE INDUSTRIAL PROCESSES

M. COEYTAUX
Serete Engineering

SECI-MANAGER is a sophisticated software package for use in optimizing production costs associated with complex energy-producing systems.

It can be employed as an investment decision making or as an equipment operation tool. As an operator aid, it is a logical addition to any modern process control system, and can improve existing control capabilities by providing a complete overview of operating events that fully reflects the interdependency of plant equipment.

The SECI-MANAGER package includes a main program designed for modeling of energy flow schemes and several interfacing functional modules for simulation, optimization and consistency calculations.

1.ENERGY SYSTEMS MODELING

Designed for use with any type of energy-producing equipment, SECI-MANAGER incorporates a powerful modeling program that enables simple, fully interactive reproduction of all energy flow diagrams

Such diagrams are constructed gradually, using "building blocks" to represent the various components of the installation. Models for these blocks are extracted from a standard equipment library of preprogrammed equations.

These equations may be in different forms of varying complexities. Many of them are non-linear, and can thus better describe the physical attributes of the equipment, to ensure a higher degree of modeling accuracy.

2. COMPUTATION MODULES

There are three such modules, all of them directly linked to the main modeling program. These modules perform three computation functions :

- simulation,
- optimization,
- measurement consistency checks.

2.1. SIMULATION MODULE

Based on :

- **plant energy requirements,**
- **cost of energy being used to satisfy these needs.**

The simulation module can compute flows, enthalpy data and capacities of all systems on the energy flow diagram, together with the total cost of production at any given time and the marginal costs of the various components.

Non-linear equations, illustrative of energy flows, are solved in a single operation, with allowance for the constraints imposed by equipment performance limits and control logic characteristics.

With this module, it has thus been possible to solve sets of more than 5,000 equations at a time.

2.2. OPTIMIZATION MODULE

Once energy needs have been defined, optimization consists of improving on a given factor, usually the total operating cost, by defining optimal setpoint values, equipment operating conditions and energy utilization schemes.

Optimization is only possible, however, where several alternatives are available, for example :
- multiple generator units for power sharing,
- choice of turbine or motor drives for rotating machinery,
- steam heater or manifold pressure adjustment capability,
- turbine generator load follow start-up and shut-down,
- deaerator temperature control.

These alternatives are determined in the simulation phase. On transfer to optimization mode, the corresponding variables are considered as unknowns and the program then determines the values leading to optimization of whatever the factor selected (cost, cogeneration efficiency, pollutant emissions, etc...).

The optimization algorithm specially developed for SECI-MANAGER software makes allowance for all of the problems involved in solving non-linear equations and in handling discrete variables, thus ensuring a high degree of computational accuracy.

2.3. CONSISTENCY CHECK MODULE

Measurement errors caused by faulty calibration, signal transmission disturbances, sensor failures, etc..., many completely jeopardize the success of the optimization phase.

To ensure the credibility of setpoint values generated by SECI software for operator guidance, it is vitally important that measured data be reliable.

The consistency check module is designed to verify that the values transmitted from the sensing devices to computer input are consistent, i.e. that they reflect reasonable mass and energy balances for whatever the scheme being analysed.

Where inconsistencies are observed, corrected values can be statistically determined for each measurement point such that, on correction :
- balance requirements are satisfied,
- total discrepancy between the corrected and the measured values, after weighing for instrumental uncertainties, is minimal.

3. APPLICATIONS FOR SECI MANAGER

SECI offers three types of computer-aided engineering capabilities. It can be used :
- in the design phase, to determine the advisability of investing in additional equipment for improving energy system efficiency or, where necessary, of fully replacing existing, but outdated installations,
- to evaluate performance of existing installations for the purpose of identifying standard operating conditions to guide plant operators in rationalizing energy purchases or establishing their energy budgets,
- as a real time operating aid, by providing the operator with a set of key variables or directly acting on controller setpoints.

Different operating conditions and various versions of the software exist,corresponding to different types of needs :
- in its OFF-LINE version, the software is located in a host-computer on the client's premises ; data are entered manually via the keyboard. Results appear on the screen or the printer or are stored on magnetic medium according to equipment configuration.

In this case, the computer is physically separated from the measurement system linked to the installation and can therefore be installed anywhere, in particular at the company's headquarters. The software is principally used for aided design, operation analysis and drawing-up forecast budgets.

- In its operator-guide version the host computer is directly connected to a data acquisition system, centralizing information coming from sensors installed on the site. The software is automatically supplied with real data. The results of optimization calculations will appear on the screen informing the operator of actions to be taken for optimum plant operation.

The operator-guide version is used for operating complex energy systems where the consequences of an operating decision are often difficult to assess.

- In its ON-LINE version data are automatically supplied to the software, as in the operator-guide version. In addition, it is capable of directly modifying equipment operating set points inside the PLCs according to the results of optimization calculations.

This version permits totally automated and optimized operation of a plant requiring no operator intervention.

4. SECI-MANAGER DEVELOPMENT

SECI-MANAGER is a living product which is continually evolving in new fields :

- Input-output where the dialogue friendliness is improved,
- The equipment library is constantly being enriched,
- Availability of functional modules.

Other developments are underway, some of which are well advanced, aimed to enable SECI to resolve industrial process simulation and optimization problems.

SERETE has already modeled a complete sugar plant.

The model took into consideration all material flows such as sugar, non-sugar, impurities, CaO, $CaCO^3$, etc... and of course energy. Corresponding balances are calculated and then optimized. This SECI sugar software is now on the market.

This SECI sugar software has been used by SERETE in several studies as an aid to optimize equipment design and configuration in the frame of extension or modification projects on existing plants.

Further developments of the same type have been launched for applications in other industrial sectors and particularly in the pulp and paper industry. A working program concerning the creation of a detailed model of the pulp plant Usine de Saint Gaudens, of the Cellulose du Rhône et d'Aquitaine (CDRA) Group is underway and being developed jointly by SERETE (Engineering Company), CTP (Technical Center for pulp and paper industry) and the Customer CDRA Usine de Saint Gaudens. The project is partially subsidized by French authorities.

The main goals of this collaboration are :

- First : in detail modeling of the pulp plant for operationnal use in the different production units in an off-line way,
- Secondly : progressive use of the program in an on-line way, in order to continuously and automatically optimize plant operation.

In this second phase, each unit will have at its disposal the SECI pulp software connected and interacting with a numerical control system. Overall plant optimization will be implemented by general plant management.

A whole model of the Saint Gaudens plant including more than 10 000 equations, has already been run satisfactorily. Our next task is to put it into operation on site. This will be completed during the first months of 1988.

TECHNICAL DESCRIPTION OF THE SOFTWARE

1. SUBJECT

This specification deals with the supply and assistance in installation and start-up of an optimized energy management software known as SECI MANAGER.

2. DESCRIPTION OF SUPPLY

The supply includes :

- SECI-MANAGER on a magnetic medium adpated to the host machine,
- Documentation in 4 copies,
- A data file for testing SECI-MANAGER.

3. TECHNICAL DESCRIPTION

3.1. DESCRIPTION

In its basic version, SECI MANAGER is used for the following :

- to simulate operation of any type of industrial facility by calculating its energy balance and the cost of power consumed,
- to seek optimum operating configuration,
- to calculate marginal costs of all the fixed variables in the system.

Two further modules may be added as an option for :

- coherence processing of measurements on site (COHER program),
- estimating the annual operating costs of the installation on the basis of operating modes and energy purchase contracts (DEX program).

Its main characteristics are as follows :

3.1.1. Simulation module

- Equiment library

SECI-MANAGER includes a library of equipment models with pre-programmed equipment operation equations for calculating operating conditions on the basis of characteristic values of each item of equipment.

Equipment models supplied include :

- utility and power connections,
- expansion - desuperheating,
- deaerators,
- flash tanks,
- back pressure turbines,
- condensation turbines,
- air condensers,
- pumps,
- exchangers with or without change in state,
- boilers (normal, electric, recovery),
- gas turbines.

Certains specific nodes are also provided :

- regulation,
- cost,
- free equation .

A thermodynamic module representing Mollier diagram of water is included in the supply.

The equipment module library is simply extended according to the characteristics of the given site.

- Diagram composition processing method

Description of the physical links between elementary equipment permits overall simulation with no limit in the number of equipment analysed (subject to the memory space available in the computer) and without pre-defined architecture.

- Taking into consideration logic operating constraints and control logic
- Non linear algorithms for solving a set of equations

SECI-MANAGER uses an algorithm for overall solving of a system of non-linear equations and in equations with no limitation as to the complexity of equations introduced.

- Easy introduction of data

Modeling any type of site and filling it from all data bases is carried out in conversational mode via a set of straightforward questions which appear on the terminal screen in simple, non-symbolic language. These questions are answered by the operator.

In particular, specific, clear questions are posed on characterics values of equipment.

- Issue of results

SECI-MANAGER offers the choice between numerous types of standard issues which range from the very detailed "step by step" to a brief summary of results.

Variables are identified :

- by a code composed of a letter followed by one or several numbers corresponding to equipment numbering,
- by clear wording.

Units are those normally used by heat engineers :

- output t/hr
- enthalpy Kcal/kg
- electricity KW
- pressure bar
- temperature °C

or I.S. units.

SECI-MANAGER allows the user to write his own sub-program for issuing results and for obtaining presentations adapted to the sites modeled (in addition to standard issues).

The simulation module is composed of 2 parts which operate separately :

- SECI 1 which is used to describe the diagram to be modeled and to fill in data bases.
- SECI 2 which does the calculation and takes into account modifications not related to the diagram.

SECI 2 calculated overall costs (in FF/hr)) according to the various tariffs (tariff periods in terms of time and season). On request, it also calculates the marginal cost of any fixed variable. For example, it can supply the marginal cost of low pressure steam (FF/t) on the set temperature of a water reheater (FF/°C) for all energy tarif periods.

3.1.2. <u>OPTIM : Optimizer module</u>

Within a previously defined optimizable field, this module seeks a configuration which offers a minimum overall cost within all the required operating limits.

This module interfaces perfectly into SECI 1 and SECI 2.

The optimization algorithm is non-linear and identifies optima which are not necessarily the mini-maxi limits. Furthermore, it deals with bivalent choices such as moto-turbo communications or post-combustion start-stop.

Any criteria may be optimized (overall cost, consumption, fuel, etc..)

3.1.3. COHER : Coherence measurement consistency module

According to the specification of all site sensors (location, type, class of precision) and values recorded on these sensors at the same time, the COHER module corrects recorded values :

- to satisfy all the diagram's mass and thermal balances,
- to minimize error according to the variance of each sensor (Gaussian distribution of differences according to the precision of each sensor).

It also assesses the quality of each sensor and indicates the gain in overall precision as a result of coherence processing.

This module interfaces perfectly with SECI 1 : any modification in the diagram is automatically taken into consideration.

3.1.4. COUTS DEX : operating costs module (optional)

This module is used to easily breakdown the year into different operating modes and to interface them with the tariff periods of the energy used, in order to calculate fixed and proportional costs of all power consumed in each operating mode and over the whole year.

In particular, this module can be used to optimize the energy contracts according to the optimal configuration obtained within each hour-seasonal tariff period.

3.2. FORM OF SUPPLY

SECI MANAGER is supplied in the form of a module which can be implemented at the time of creation of any equipment defined by the user.

It will be delivered and tested with a test game.

3.3. DOCUMENTATION

3.3.1. General documentation

- Methodology,
- Description of equipment nodes,
- Description of special nodes (control, etc..),
- Procedure for drawing a diagram of a site,
- Preparation of data bases for the creation of a structure.

3.3.2. SECI 1 and SECI 2 operating manuals

- Detailed explaination of all menu options,
- Interpretation of error messages.

3.3.3. Listing of equations

- Equations of the equipment in the library

4. TEST GAME

The test game is designed to verify the identity of results between SECI MANAGER operating on the supplier's premises and SECI MANAGER at the client's premises.

5. INSTALLATION CONDITIONS

In view of the language used (FORTRAN 66 compatible 77), SECI is a portable software which can be used on a very wide range of hardware.

- main memory available : about 3 Mbytes
- hard disk : 40 Mbytes or more
- calculating speed : APPROX 1 Mips
- Architecture 32 bits.

The above are not intangible values but correspond to average characteristics for operating SECI software under optimum conditions.

ENERGY SAVINGS AND ECONOMIC CONSEQUENCES RESULTING FROM THE INSTALLATION OF A COGENERATION UNIT (ELECTRICITY - STEAM) AT THE CORINTH REFINERY - (MOTOR OIL HELLAS)

A. KALYVAS
MOTOR OIL (HELLAS) CORINTH REFINERIES S.A.

1. INTRODUCTION

The MOTOR OIL Refinery in Corinth is a complex Refinery both with respect to the various processes and techniques used and with respect to the range and quality of products produced. It produces the whole range of oil products and lube oils and employes various techniques such as Atmospheric Vacuum Distillation,Reformation,Fluid Catalytic Cracking (FCC),Alkylation etc.

For the optimum guidance and coordination of the refining operations, linear programming techniques are used with the aid of an IBM 4331/IBM computer.

In order to cover its needs in electricity the Refinery has been using the Public Power Corporations' (PPC) network.

The needs in steam for the different refining processes used in the Refinery,were covered by the steam boilers' production using liquid or gas fuels.

Refinery operations in recent years caused an excess production of fuel gases which were burnt at the "flare".The decision to increase the Fluid Catalytic Cracking (FCC) processes' capacity contributed to an increase of excess fuel gases being burned at the "flare".

Under this context a feasibility study was undertaken in 1983 in order to examine the possibility of using part of the excess fuel gases for the production of electricity. The conclusions of this study,which also examined the additional costs due to the frequent power cuts in the PPC's network convinced the companies' management to immediately undertake the execution of the project.

The project had the financial support of the European Economic Community (DEMONSTRATION PROJECT IN THE FIELD OF ENERGY SAVINGS),as well as that of the Greek State in addition to its financing by the European Investment Bank,the Credit Lyonnais and the National Investment Bank of Industrial Development.

2. TECHNICAL CHARACTERISTIQUES OF THE INSTALLATION

The main installation characteristics are the following:
Fuel gas compression system and LPG vaporisation system
operating pressure : 3,5 Kg/cm^2
operating temperature : 35° C
molecular weight : 20.2 - 16.3
after the compression to 16,4 Kg/cm^2 the vaporised LPG has a pressure of 14,3 Kg/cm^2 at 95° C
Gas turbines and alternators
2 turbines and 2 alternators of 11.5 MW each at 35° C at sea level pressure.

Waste heat recovery and steam production unit :
High pressure steam
production : 52 ton/hr
operating pressure : 48 Kg/cm^2
operating temperature : 420° C
Low pressure steam
production : 16 ton/hr
operating pressure : 2,5 Kg/cm^2
operating temperature : 138° C
The installation made provisions for the coverage of 97,8% of the Refineries' electrical needs.

3. PRODUCTION AND MEASUREMENTS

The studies,construction works,and trial runs of the installation were completed by October 1984.Normal production started in Nov.1984.Systematic measurements were undertaken between Nov.1984 and 1985 in order to make the evaluation of results possible.

New parameters were introduced and were considered in the linear programming,in order to optimise total production (production of the Refinery and of the Cogeneration Unit)and especially the production and quality of the gases particularly those produced by the FCC and Reformation units.

All measurements taken as well as the measurement results concerning the gas composition are shown on tables I to III.

4. PRODUCTION RESULTS

The theoretical foundations for certain calculations and evaluations are from the publication "THERMODYNAMIQUE ET ENERGETIQUE" by Professor L.Borel of the Ecole Polytechnique Federale(EPFL)of Lausanne and from the article "EFFICACITE ET RENDEMENT EXERGETIQUE D'UNE PETITE CENTRALE D'ELECTRICITE-CHALEUR" by A.Tastavi of the "Laboratoire de Thermodynamique"of the "EPFL"(Bulletin Technique de la Suisse Romande 24 Juin 1982).

In addition,the extreme case when there is no reduction in the waste of gases to the"flare"($\dot{L}_F$ = 0) was calculated.

Figures 1 and 2 represent and illustrate the evaluation of different magnitudes for the period examined.The main results are the following :

a) Total Production for the period Nov.84-Oct.85

	84-85	INITIAL ESTIMATE
-Electricity production		
10^6 Kwh/y	151,5	174,4
Coverage of needs	94,3%	96%
-Steam production		
10^3 MT/y	457	544
Coverage of needs	51,3%	66,7%
-Gas consumption		
10^3 MT/y	51,8	68,5
Reduction of waste		
to the "flare"(gas) 10^3 MT/y	11,04	11,0

b) Results and energy saving

-Energy efficiency	75,64%	75,91%
-Usable energy efficiency	43,95%	42,63%

-Energy savings at the Refinery level MTOE/y	9139	8418
-Usable energy savings at the Refinery level MTOE/y	36923	37931
-Energy savings at National level MTOE/y	36229	37345
-Usable energy savings at National level MTOE/y	36668	37182
c) Commercial Results		
-Electricity cost $/1000 Kwh	43,95	45,02
-Reduction in the cost of steam $/MT	1,76	2,68
-Annual profits of the Refinery 10^3 $	6489	6306
-Investment yield	33,06%	32,04%
1* Estimation of gains at the National Economy Level 10^3 $	4999	6132

* Note 1

We are only calculating the economies from the combustible liquid used in the production of electricity of a thermal origin by the Public Corporation.

5. CONCLUSIONS

It can be said without doubt that the Cogeneration investment in the Corinth Refinery presents impressive results both in the saving of energy and in the economic returns under the market operating conditions of the years 84-85. It would be interesting to study the extreme operating conditions and the generalisation of the results. The tables,figures and results drawn from the project make the study of every case possible as well as the derivation of conclusions.

We list the most important factors which must be considered. These are :

a) Waste from the "flare"

b) Price of Fuel Oil

c) Price of electricity purchased from Public Corporation

In the extreme case when it is not possible to reduce the waste from the "flare" ($\dot{L}_F = 0$),the yield of the investment is still interesting.

A) Without reduction in the waste from the "flare"

	Price of Purchased Electricity $/1000 Kwh	
-Refinery Level	30-40	50-60
Fuel oil price $/MT	< 90	190
-2*National Level		
Fuel Oil price $/MT	> 110	110

*Note 2 See Note 1

B) With reduction in the waste from the "flare" at the level of 11.000 MT/y

-Refinery Level		
Fuel Oil price $/MT	< 180	430
-National Level		
Fuel Oil price $/MT	> 70	70

TABLE 1

Symbol	Units	Design	Flow rate at flare 0 Nov. 84	Dec. 84	Jan. 85	Feb. 85	Mar. 85	Apr. 85	Enthalpy 02,03:128 May 85	Jun. 85	July 85	Aug. 85	Entropy 02,03:382 Sep. 85	Oct. 85	Average
$\dot{M}_{1v}$	T/h	34·000	62·367	44·024	46·774	54·250	55·361	39·847	44·619	45·986	43·653	45·753	47·124	71·647	50·117
$\dot{M}_{2v}$	T/h	52·000	25·574	45·750	39·699	39·139	43·431	42·656	43·817	41·931	45·072	44·843	41·786	15·031	39·061
$\dot{M}_{3v}$	T/h	16·000	8·257	14·975	15·524	14·299	16·038	15·418	15·815	14·932	15·257	15·588	14·614	5·264	13·832
$\dot{M}_{L}$	T/h	0·000	0·000	0·000	0·000	0·000	0·000	0·000	0·000	0·000	0·000	0·000	0·000	0·000	0·000
$\dot{L}_{F}$	T/h	1·375	1·526	1·621	1·042	0·896	1·341	1·465	1·532	1·372	1·425	1·561	1·248	0·303	1·278
$\dot{M}_{FG}$	T/h	8·568	4·979	6·300	6·031	5·799	6·488	6·204	6·922	6·919	7·365	7·090	5·931	1·931	5·997
$\dot{M}_{1FG}$	T/h	1·318	0·601	0·119	0·040	0·040	0·000	0·000	0·489	0·688	0·744	0·503	0·049	0·007	0·273
$\dot{M}_{FO}$	T/h	1·187	4·126	3·272	3·567	4·147	4·278	3·094	2·888	2·779	2·529	2·944	3·582	5·513	3·560
$\dot{H}_{IFG}$	10^3 kcal/T	11 000	11 304	10 703	10 834	10 418	11 500	11 488	11 672	11 305	11 125	11 427	11 720	10 987	11 207
T_1	°C	390	375	370	360	370	370	360	370	370	370	360	350	380	367
T_2	°C	420	405	405	405	405	405	405	405	405	405	405	405	405	405
T_3	°C	138	143	143	143	143	143	143	143	143	143	143	143	143	143
P_1	kg/cm²	47·0	45·0	45·0	45·0	45·0	45·0	45·0	46·0	46·0	46·0	46·0	46·0	46·0	45·5
P_2	kg/cm²	48·0	48·0	48·0	48·0	48·0	48·0	48·0	48·0	46·0	46·0	46·0	46·0	46·0	47·2
P_3	kg/cm²	2·5	3·2	3·2	3·2	3·2	3·2	3·2	3·2	3·2	3·2	3·2	3·2	3·2	3·2
$\dot{T}_a$	°C	25	16	13	12	11	13	19	24	28	29	29	26	20	20
$\dot{E}$	10^3 kcal/h	18 748	17 824	16 397	15 806	15 407	18 144	17 518	18 189	16 662	18 823	19 313	16 781	4 606	16 289
$\dot{E}_{PPC}$	10^3 kcal/h	774	4 778	1 292	161	306	106	76	76	306	497	268	101	3 860	986

TABLE 2

Symbol	*Units*	*Measurements*
$\dot{M}_{1v}$	T/h	Weight flow of steam in state 1 (generator outlet)
$\dot{M}_{2v}$	T/h	Weight flow of steam in state 2 (cogeneration installation outlet)
$\dot{M}_{3v}$	T/h	Weight flow of steam in state 3 (cogeneration installation outlet)
$\dot{M}_L$	T/h	Weight flow of gas at flare
$\dot{L}_F$	T/h	Gas loss at flare
$\dot{M}_{FG}$	T/h	Fuel flow rate (gas)
$\dot{M}_{1FG}$	T/h	Fuel flow rate at boiler (gas)
$\dot{M}_{FO}$	T/h	Fuel flow rate at boiler (fuel oil)
H_{lFG}	10^3 kcal/h	Lower energy output of fuel (gas)
T_1	°C	Temperature in state 1
T_2	°C	Temperature in state 2
T_3	°C	Temperature in state 3
P_1	kg/cm^2	Pressure in state 1
P_2	kg/cm^2	Pressure in state 2
P_3	kg/cm^2	Pressure in state 3
T_a	°C	Ambient temperature
$\dot{E}$	10^3 kcal/H	Electrical output of turbines
$\dot{E}_{PPC}$	10^3 kcal/h	Power received by electricity company
h_1	10^3 kcal/h	Steam enthalpy in state 1
h_2	10^3 kcal/h	Steam enthalpy in state 2
h_3	10^3 kcal/h	Steam enthalpy in state 3
h_{01}	10^3 kcal/h	Water enthalpy at boiler inlet—entry to cogeneration installation (states 2, 3 = 128)
s_1	kcal/T	Steam entropy in state 1
s_2	kcal/T	Steam entropy in state 2
s_3	kcal/T	Steam entropy in state 3
s_{01}	kcal/T	Water entropy at boiler inlet—entry to cogeneration installation (states 2, 3 = 382)

Table III

COMPOSITION OF GAS

Component mole %	NOV. 84	DEC.84	JAN.85	FEBR.85	MARCH 85	APRIL 85	MAY 85	JUNE 85	JULY 85	AUG.85	SEPT.85	OCT.85
H2	35.4900	23.2700	27.1400	5.4810	40.4500	41.9830	42.4000	38.9720	23.6000	33.2400	45.1900	14.1800
C1	18.2800	22.3300	18.2500	14.6220	17.9500	16.3900	17.2000	15.8570	19.5000	18.7900	13.9400	4.1900
C2	7.7900	7.4200	11.7400	8.3620	8.9900	11.8650	12.1700	14.0170	15.9500	13.7000	11.9500	3.7700
C2=	5.7200	5.6800	7.0100	6.7130	6.3000	7.3250	6.6200	7.5580	8.3400	7.1700	4.4300	1900
C3	12.5000	17.4000	11.9500	16.8180	10.6300	7.4520	7.8000	7.4790	17.4000	12.6200	12.8300	37.0200
C3=	9.9300	6.5700	10.8000	27.5520	8.9400	7.3940	4.6000	5.2890	7.1000	6.4200	3.5400	13.2400
i-D4	2.3600	4.9800	2.3600	3.7050	1.1400	1.7190	2.2000	2.9690	1.9000	1.7400	1.9800	6.1400
n-D4	2.5800	4.8400	2.1500	3.4590	2.0600	1.8870	3.1000	2.7690	2.2500	2.7300	3.3400	15.9500
1-D4=	1.0000	0.5900	1.1500	1.9920	0.4000	0.4020	0.7400	0.5100	0.2300	0.4000	0.2500	1.1700
c-D4=	0.1900	0.2500	0.8200	1.2910	0.2100	0.2070	0.4800	0.2800	0.1000	0.2100	0.0900	1.1400
t-D4=	1.1000	0.1600	0.7500	1.4070	0.1800	0.1940	0.4900	0.2700	0.0800	0.1200	0.0900	0.9100
i-D5	0.0000	0.0000	0.1300	0.4090	0.0800	0.0370	0.1700	0.2600	0.0800	0.4600	0.3300	0.0500
n-D5	0.0000	0.0000	0.0000	0.1250	0.0000	0.0330	0.0700	0.0500	0.0600	0.1500	0.2000	0.1600
C5=	0.0000	0.0000	0.0000	0.0000	0.0000	0.0000	0.0000	0.0000	0.0000	0.0000	0.0000	0.0000
CO	0.1400	0.2600	0.1000	0.1050	0.1000	0.0970	0.1000	0.1800	0.1400	0.1400	0.1000	0.0800
CO2	0.0500	0.1400	0.0400	0.0430	0.0500	0.0700	0.0600	0.0600	0.0700	0.0700	0.0600	0.0400
N2	2.8700	6.1100	5.6100	7.9140	2.5200	2.9450	1.8000	3.4790	3.2000	2.0400	1.6800	1.7700

MOTOR OIL HELLAS Steam/electricity cogeneration installation

Fig.1

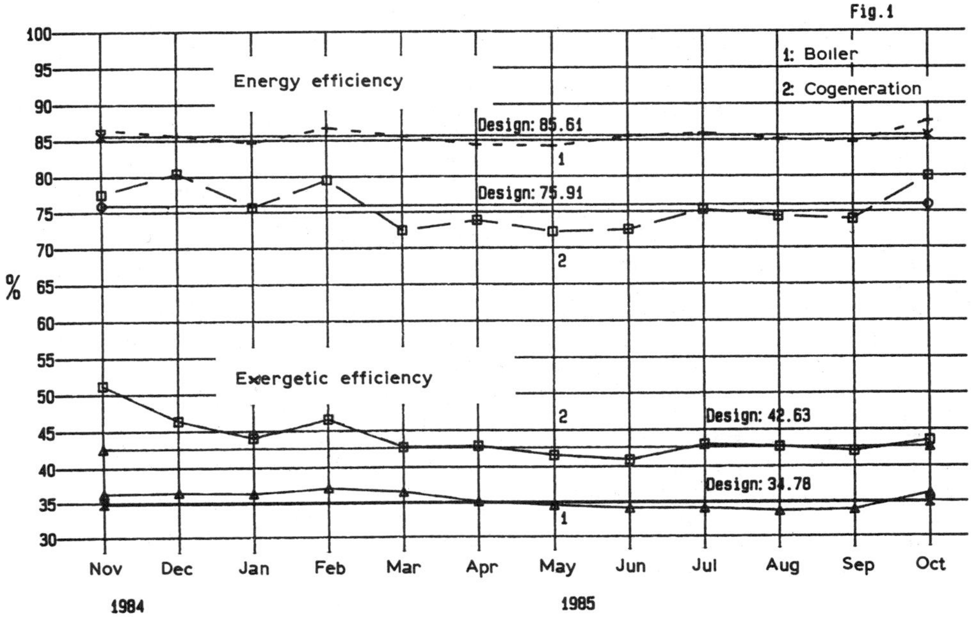

MOTOR OIL Steam/electricity cogeneration

Fig. 2

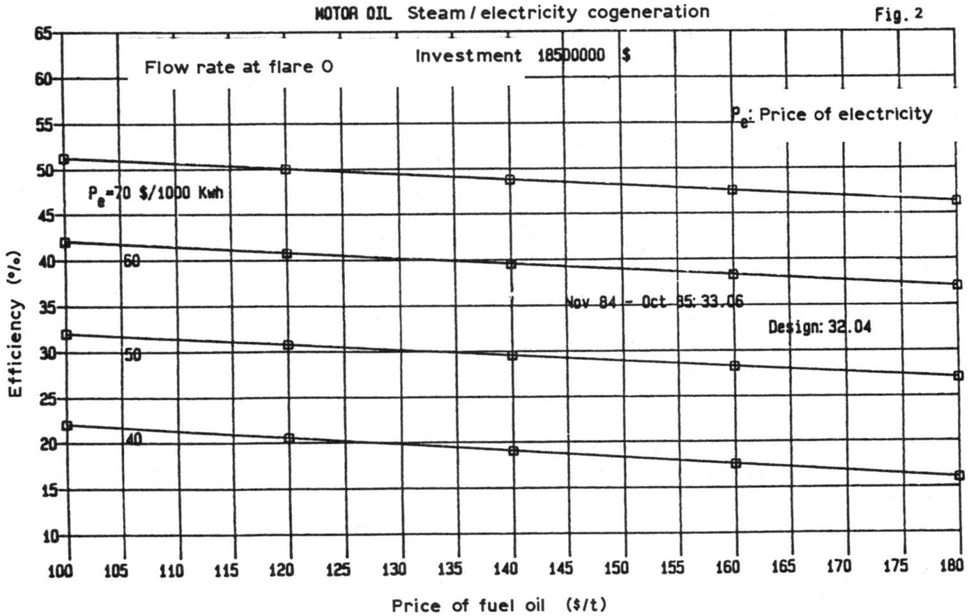

SOFTWARE SYSTEMS TO OPTIMIZE COMBINED HEAT AND POWER PLANT

J SPRINGELL & D FOSTER
IMPERIAL CHEMICAL INDUSTRIES PLC
ENGLAND

ABSTRACT

Selecting the best way to run a large industrial Combined Heat & Power systems to meet site electricity and steam demands at minimum cost is a highly complex problem. With typical annual fuel bills of several millions of pounds, even small percentage changes in operating efficiency can have a large cost effect.

An optimisation system which incorporates a detailed power plant simulation model with an appropriate mathematical optimisation algorithm can determine the optimum settings for control variables of the power plant and eliminate uncertainties associated with achieving the minimum cost operation.

The Control & Electrical Group of ICI's Engineering Department have developed a system called HAMBLE to achieve these optimization objectives. Both on-line and off-line systems have been installed. These systems can solve the optimization problem quickly and in real time, yielding operating costs savings of two to five percent with short paybacks.

An overview of these systems is presented. Their application is illustrated by a case study.

1. INTRODUCTION

Production overcapacity in the western world, weak demand and squeezed profit margins are among the factors that have forced Imperial Chemical Industries (ICI PLC) to improve the operating economics of their existing chemical plants. The company has tackled this problem by integrating its production management experience, process technology, and cost information and has incorporated these into plant computer systems. Such systems are capable of assessing plant operating economics and advising operators or automatically initiating changes in running conditions which would be economically beneficial. The systems have been developed over a 20 year period - culminating with real time data collection and performance monitoring (Ref 1) and real time optimisation for a variety of plants, including heat and power plants (Ref 2). ICI has found that they have been able to use these systems to generate significant improvements in efficiency and profit margin at low capital expense.

The basic concept has been successfully applied to a variety of different plants with particularly good results being achieved on Combined Heat and Power Plants. The first CHP optimiser system went on-line in 1981 and, since that time, ICI has installed a similar system in each of their sites.

The concept is simple, yet highly effective. It involves using economic information (i.e. marginal costs) coupled with operating plant data (i.e. temperatures, pressures and flows) and mathematical models to help plant personnel to make decisions on how best to operate the plant.

By constructing a rigorous simulation model of the plant, it is possible to predict plant operating conditions from a given set of critical parameters (e.g. flows, temperatures, pressures) and to investigate the effects of making changes to these parameters. By incorporating cost information into the simulation model and applying a fast-acting mathematical optimisation technique it is possible to determine the set of optimal operating conditions which in the case of heat and power plants meet the site demand for steam and electricity at minimum cost. In addition the system reveals the marginal cost of an incremental increase in steam or electrical output and this provides useful information upon which the consuming plants can base operating decisions. ICI's experience in the application of these techniques to industrial CHP plants is that they typically reduce plant operating cost by 2% to 5% while fully meeting site demand.

In terms of practical power plant operation, steady state operation is a condition that is achieved only very occasionally. Ambient conditions, electrical tariffs and site steam demands change frequently and so a fast response is necessary if minimum cost operation is to be maintained at all times.

ICI has resolved these problems by building a fast-acting model and optimiser which performs economic optimisation. The model is linked with database software and interfaced with plant instrumentation resulting in an on-line system that solves the optimisation problem in real time.

2. DESCRIPTION OF MODEL/OPTIMISER SYSTEM

The daily problem faced by Combined Heat and Power Station operating personnel is to decide the least cost way to meet the site energy demands within the internal constraints on operation imposed by the power station itself. Should the in-house electrical generation meet the site demand for electricity exactly or should the grid import/export facility be involved? What should be the loadings on various boilers, gas turbines and steam turbines in view of the non-linear changes in efficiency with loading? Is it worth venting LP steam to increase power generation? All of these questions and more need to be answered to provide an effective solution.

Typically, the independent control variables in these types of problems are such things as boiler delivery rates and temperatures, turbine throttle flow rates, condenser rates, LP vent rates, desuperheater flows, etc. In the context of optimisation, these variables are known as the degrees of freedom of the system. Limits on the ranges within which the degrees of freedom must lie are called simple inequality constraints i.e. minimum and maximum boiler rates, minimum and maximum turbine throttle rates,etc. More complex inequality constraints originate from the physical limits of process operation. The degrees of freedom can then be manipulated within all their constraints to meet the site energy demands which, in the context of optimisation, become equality constraints and must be met exactly. The final choice for the values of the independent controlled variables is determined by the set of values that give the lowest operating cost subject to the inequality and equality constraints. For optimisation of CHP plants, operating cost is the objective function that must be minimised.

3. THE MATHEMATICAL MODEL

The first step in ICI's approach to the solution of this strategic problem has been to develop a detailed heat, work and mass balance model of the power station. This model provides an accurate representation of steady state performance. Given a set of values for the degrees of freedom it will predict a complete energy flowsheet of the station including the steam exports, in-house electrical generation and overall cost. This is shown in fig.1.

Within the model, each boiler, turbine and heat exchanger has a separate non-linear mathematical representation which has been matched to its physical counterpart. The non-linearity in some of the descriptive equations is very relevant to the operating decisions and makes the mathematical description of the system very accurate. The model must also account for the thermodynamic interactions between individual items of equipment and therefore must include rigorous steam and water physical property routines.

Using the mathematical model, the plant engineer can quickly calculate the overall cost and levels of steam export from the power station. Hence he can investigate the way in which these important quantities vary with changes to internal operation. By this means he acquires a greater understanding of how the station behaves as a whole - not a trivial task for a large integrated system. The use of the model to calculate the steam exports from reliable power readings has proved more accurate and consistent than the traditional way which was based entirely on flow measurements. The frequent running of the model forms the basis of a computer-based 'historical data' record system where averages of important calculated quantities, including steam exports and cost, are logged every five minutes. The five minute data is averaged hourly to generate an entry into a detailed record of operation which is kept for three years.

Although the model does provide a tool with which to investigate how steam demands depend upon the free variables of the power plant, it is inconvenient to adjust the selected variables manually to achieve the desired steam and power demands. This problem is complicated by the simultaneous need to find the minimum cost solution. This task is addressed by a fast-acting numerical optimisation algorithm which can choose values for the set of independent variables that simultaneously meet the demands of the site at minimum cost.

4. OPTIMISATION

The second step in the ICI approach to solve the CHP optimisation problem involves applying a mathematical technique for non-linear constrained programming.

Normally, solution of this type of problem requires that the model equations be solved each time that the optimiser algorithm generates a set of values for the independent variables. This is very inefficient. Ideally, the model equations should be solved only once, and that is when the optimum set of values for the independent variables has been found. Treating the model equations as equality constraints in the optimisation allows the algorithm to simultaneously converge to the solution of the optimisation problem and the mathematical model. This 'one shot' formulation of the problem increases the number of optimisation variables, but with the right optimisation algorithm the improvement in performance is impressive.

This approach has allowed ICI to design a system that can find the minimum cost solution to an operating scenario in a few minutes of CPU time, resulting in a system which can run on-line and in real time. This ability to run on line in real time makes it an important tool to aid station operations staff to run more effectively .

5. CASE STUDY - THE WILTON POWER STATION

The Wilton Power Station, located in ICI's Production facilities in Teesside, England, provides steam to the production units at three pressure levels (900, 250 and 20 psig). The power station also generates electricity which partly or fully meets the electrical demands of the site, depending on the tariff that is in force at the time. Typically, the total steam export from the power station is of the order of 1.8 million pounds/hour and the electrical generation is of the order of 200 MW. The Combined Heat and Power (CHP) Station is shown in its industrial context in Figure 2.

The Wilton Power Station has five boilers that raise steam at 1700 psig and an additional three boilers that raise steam at 900 psig. There is a capability to let 1700 psig steam down to 900 psig, 900 psig to 250 psig and 250 psig to 20 psig. The electrical generation is provided by 11 turbo-alternators. Seven are primary turbines which exhaust at 250 psig and four are secondary turbines, two of which exhaust to surface condensers and two of which exhaust to vacuum deaerators in the 1700 psig feedwater heating train. Any imbalance in electrical generation is made up by invoking the import/export facility from the National Grid. A low pressure steam venting facility allows increased use of turbo-alternators in cases where steam demand is low. Figure 3 shows a schematic diagram of the Wilton Power Station.

Additional complexity is added to the operations' decision-making process by the nature of the boiler feedwater preheating trains and the manner in which they interact with the secondary turbines. Figure 4 shows a typical 1700 psig feedwater preheating train. The make-up water is preheated to boiler inlet temperature by steam at four levels in a series of heat exchangers. Finally, all the steam that was used for preheating is condensed into the make-up water. Approximately 0.4 pounds of condensed steam is added to every one pound of fresh make-up water.

It is obvious from the diagram that the high degree of complexity and interaction of the system makes an optimisation program a must. Additional complexity is added by the ability of the boilers to be fired by either fuel oil or gas and the variation in electrical tariffs between day and night.

Since the late 1960s, the operation of the power station had been optimised by a relatively simple batch computer program which was run off-line. By the late 1970s it was becoming obvious that there was a need for a more refined optimisation program that could be run continuously on-line using real time data to reflect the true operating point of the power station.

In March 1982, the HAMBLE program was commissioned. The major objectives of the program were:

1. To continuously provide the minimum cost strategy of operating the station for the given equipment on-line.
2. To monitor actual performance against this minimum cost strategy
3. To use HAMBLE in 'what if' mode to evaluate possible changes in on-line equipment.

The current operating point is determined by data collected on-line from plant instruments. Information pertaining to the fuel and electricity costs, equipment on-line and other equipment constraints is entered by the operators through the control room terminal.

The mathematical model is run every few minutes to calculate the actual demands and associated costs of production. The optimisation program then uses the mathematical model to calculate the minimum cost solution to the operating problem. Thus the computer continuously monitors achieved actual cost and compares it to the practically achievable minimum cost solution, providing a clear incentive to efficient operation. The system also reveals the cost benefit of relaxing the binding constraints and the marginal cost of incremental changes in energy demand.

The following typifies the sort of decisions which need to be taken. Day by day the first decision is what boilers, turbines or degraders to have on-line to meet the actual and forecast energy demand. Where demand is falling the question might be whether to take a 1700 psi boiler off-line and put a smaller 900 psi boiler on-line. With a 900 psi boiler on-line it can be run at minimum load through the 900/250 degrader or at a higher load sufficient to run primary 1 turbo-alternator.

The problem is further complicated with the option of which secondary machines to run. Finally, since day electricity import from the National Grid is expensive and night import cheaper, the decision as to what equipment to have on-line has to take into account the changing tariffs.

These decisions are made using the HAMBLE model and optimiser as a calculating tool or in 'what if?' mode. This is a facility whereby the control room personnel can quickly set up, through the VDU, comparative optimised flowsheets around the current operating position, to compare by inspection the changes in costs involved.

It is quite usual for as many as 40 'what if?' optimised flowsheets to be run by the control room personnel each day.

6. RESULTS

Upon commissioning of the HAMBLE system, the operating costs of the Wilton Power Station were reduced by a full 2.5% from an already well optimised situation. Half of this savings can be attributed to better solution to the problem of how to load equipment that is on-line. The other half of the savings can be attributed to better selection of what equipment to have on-line at any given time.

7. CONCLUSIONS

The dramatic savings that were achieved on the Wilton Power Station exemplify the tremendous savings that can be achieved, even on well-run Combined Heat and Power Plants, by the installation of a model/optimiser system. This type of system can help reduce operating costs by 2% to 5% and pay for itself in less than a year.

8. REFERENCES

(1) SPRINGELL, J. "Coaudination a System for Co-ordinating an Energy Management Policy". Proc.Instn.Mech.Engrs.May 1985.

(2) FOSTER, D. "Economic Performance Optimisation of a Combined Heat & Power Station" Proc.Instn.Mech Engrs. Vol.201 No A3, 4 February 1987

(3) RANADE, S.M. and W.E. ROBERT, "Marginal Utility Cost Concept Maximises Plant Efficiency," To be published in Power Magazine, July 1987.

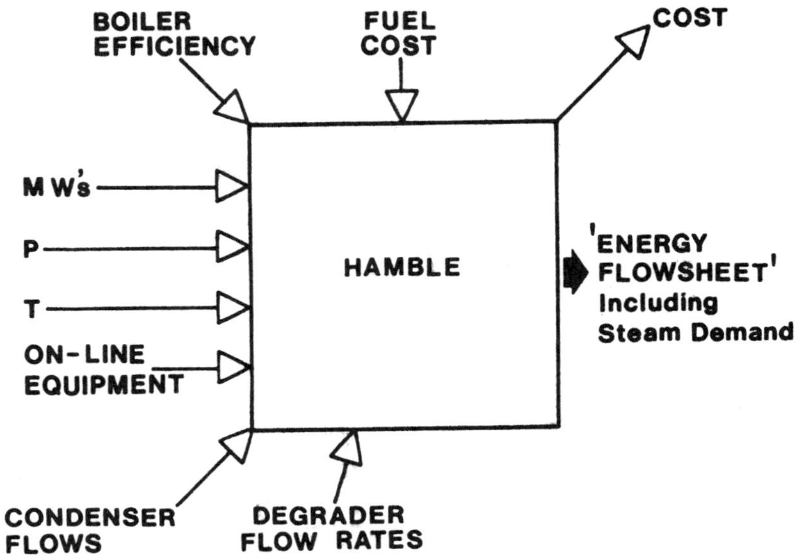

Fig. 1 Heat Work and Mass Balance Steady State Simulation (HAMBLE)

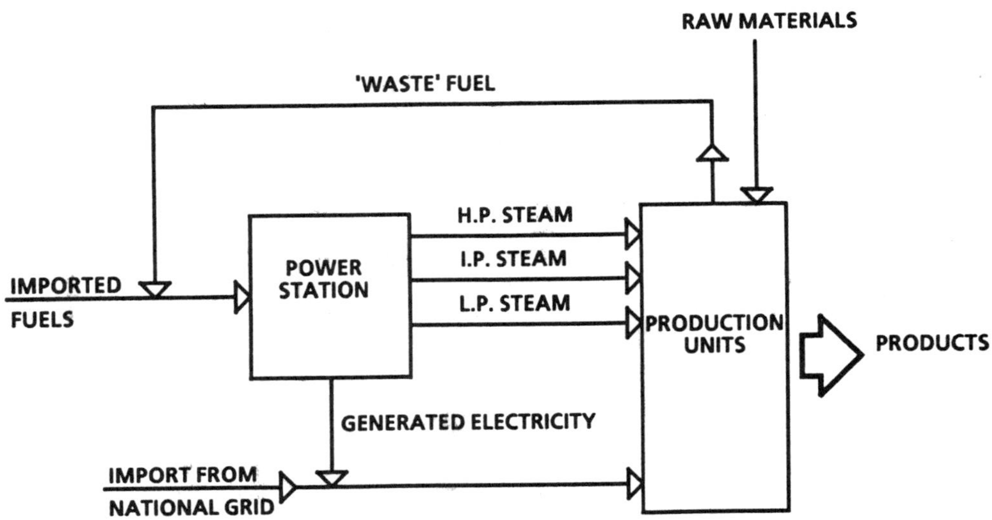

Fig. 2 A CHP Station in the Industrial Context

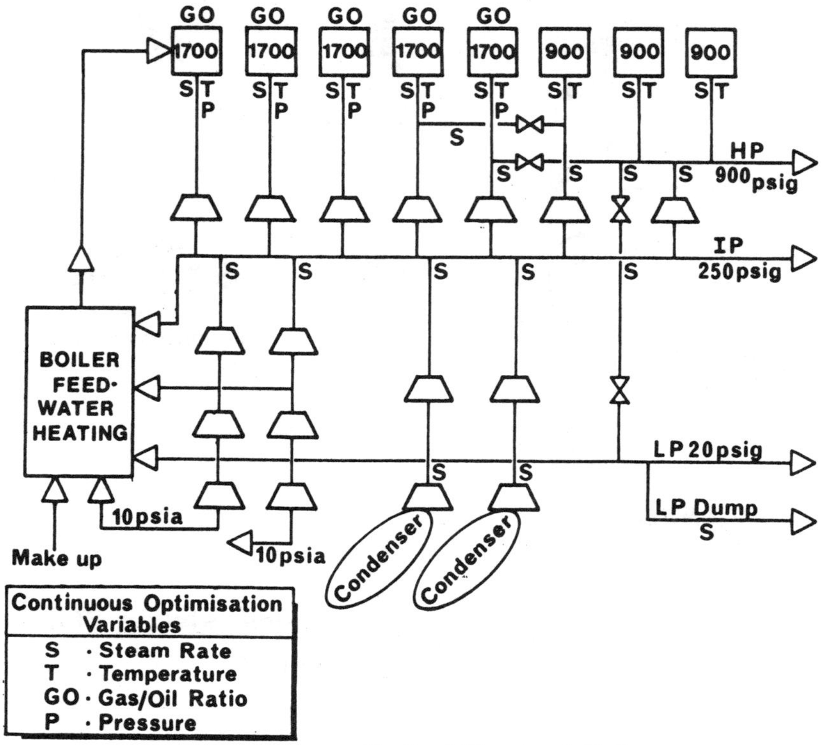

Fig. 3 Wilton Power Station

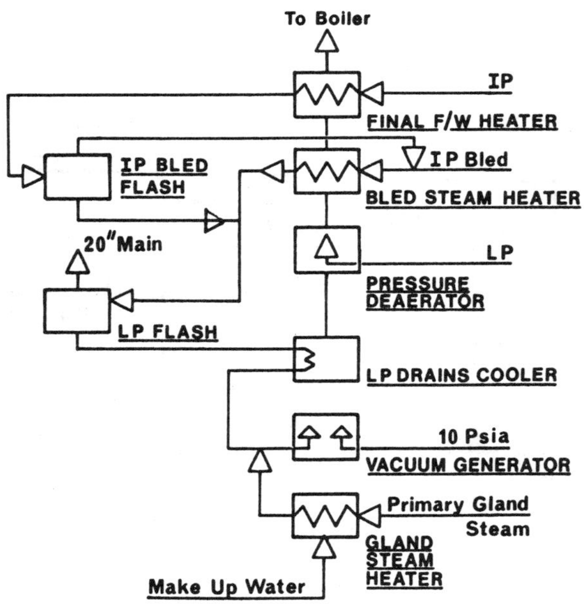

Fig. 4 1700 Boiler Feed Heating Train

SESSION VI : ROUND TABLE ON FINANCING ENERGY EFFICIENCY INVESTMENTS

The financial engineering activity of the Commission

Accelerating discrete energy efficiency investments through third party financing

A new source of finance for investments in energy savings

Financing investment in energy efficiency from the manufacturer's point of view

THE FINANCIAL ENGINEERING ACTIVITY OF THE COMMISSION

H. Carré and W. Faber
Directorate General for Economic and Financial Affairs
Commission of the European Communities

Being subject to considerable budgetary constraints, the Community is constantly considering alternative and modern approches to achieve its aims. As part of this process to find more efficient policy tools, the Commission in recent years has developed its financial engineering activities. In so doing, it is seeking to make the resources available better suited to the real needs of European economies.

Financial engineering is thus not an end in itself. Its purpose is to help attain the Community's major objectives : unifying the internal market, stimulating technological progress and job creation, improving industrial competitiveness and integrating the outlying regions. A more rational utilization of energy must certainly be considered as help to achieve those major objectives.

The framework for financial engineering is relatively favourable because private capital in Europe is plentiful. However, too often it is invested in purely financial operations that are self-perpetuating and self-propagating with speculation their driving force. We need to reverse this trend and to channel European capital into financing the investment which the Community needs.

In this context, three priority needs are currently at the centre of the Commission's preoccupations : reviving the spirit of enterprise, industrial renewal and strengthening the Community's basic facilities. These three basic needs are reflected in the Commission's approaches to financial engineering.

Objective I : Strengthening the EC's basic facilities

The Commission's focus here are large-scale infrastructure projects of European interest, for which a large amount of capital has to be assembled in special forms and using special procedures.

The Commission intends to call for a re-distribution of roles in the promotion, financing and management of such projects and for an improvement in the environment of private capital. The new role that the Community could play will be threefold:

1. The Commission intends to provide the requisite conditions for large-scale projects to emerge and be launched. This shall be done by making a budget contribution towards the cost of the necessary preparatory work, by issuing a "declaration of European interest" that would publicly demonstrate its support for a project; and by making a budget contribution at the seed capital stage, that would exert a leverage effect and would make it easier to raise the necessary volume of equity capital;

2. Furthermore, the Commission plans to improve the environment for private investors : in order to make securities issued by bodies carrying out large-scale projects more attractive they would have to be able to circulate freely within the Community and would have to be accorded appropriate tax treatment;
3. Last but not least, the Commission plans to mobilise the market through a new form of Community assistance by using the Community budget to give the European Investment Bank a guarantee on its loans. The guarantees would be in a form tailored to the risks involved, and the loans would be made either out of the EIB's own resources or from those borrowed by the Commission. The Community would thus become involved in "project financing" techniques : these have enabled a number of large-scale projects to be implemented by sharing the risks among the parties to the financial package.

Objective II : Promotion of industrial renewal

The Commission is equally active in the promotion of industrial renewal through the financing of high technologies.

Equity capital is the best form of finance for projects that are situated midway between research and industrial application. Yet, the provision of equity capital is particularly difficult to organize if the project is the result of international cooperation or if it is a long way back in the chain which runs from research to industrial application. Wide-ranging joint research programmes are however being part-financed by the Community or undertaken with its collaboration within a broader framework such as European Research Coordination Agency or "EUREKA". The projects making these programmes are at the research stage or at the pre-competitive development stage, for which grants are the appropriate form of finance since their success is too uncertain. Against this, no specific mechanism exists for projects that represent the industrial follow-up to those programmes.

To overcome this particular difficulty, the Commission has sounded out the financial community and professional circles and has found confirmation of the relevance of new financial packages. These would be based on the setting up of investment companies, with exclusively private capital, and eventually the establishment of a guarantee scheme for equity investments in high technology transnational projects.

Objective III : Reviving the spirit of enterprise

The aim to revive the spirit of enterprise is covered by the Community's programme to provide small and medium enterprises with better finance.

The Commission has explored a number of new approaches to this problem. A few years ago, it embarked upon some decidely innovative measures : the financing of innovation and the development of a genuinely European venture capital activity.

The task will be to continue and expand those measures, which are aimed : at increasing the supply of funds lent to finance capital investment of small and medium-sized enteprises; at providing access to credit for firms whose financial standing is too low to provide the necessary security (development of mutual guarantee machinery); at increasing the equity capital of SMEs (promotion of European venture capital) or at providing them with the financial and other services they need in order to grow. The creation by the Institutions of the European Community specialising in long-term credit of a European Financial Engineering Company heralds a first success for this initiative.

Financial engineering and energy saving investment

As was mentioned before, financial engineering must make it possible to satisfy funding needs that are not satisfactorily met by the market.

With this basic rule in mind, one can analyse a possible contribution of the financial engineering techniques developed and promoted by the Commission to the financement of energy-saving investments.

One of the main reasons for the slow development of energy-saving investments is the lack of companies offering their expertise and their services to reduce energy consumption, the lack of Energy Service Companies or ESCOs. The reasons for this have to be analysed. One of the causes for the reticence to set up ESCOs even though they are expected to make profits seems to be a lack of the appropriate form of finance needed for their foundation.

The amounts of money to be invested in energy-saving projects are considerable and the installations will often be specific to the project. The capital is thus immobilised in a specific project and will lose its value to a substantial degree if the project concerned fails. Thus, the capital invested to save energy in someone else's premises will only to a very small extent be considered as security for loans. We would see this to be a structural financing problem of ESCOs but which does however not differ very much from problems occurring for example in leasing companies specialised in the leasing of investment goods adapted to the needs of one specific company.

The resulting difficulties are made more acute by the problem of gearing, that is to say the debt to equity ratio. In effect, every time the ESC0 successfully installs a new piece of plant and receives a stream of shared profits in payment, it will have to borrow the funds to pay for the costsof the installation. Therefore, the more successful the ESCO, the greater the debt and the level of the ECSCO's capital becomes smaller and smaller in comparison. In these circumstances, it would only take a small reduction in the revenues to result in the ESCO having difficulties in servicing its debt.

The reticence to provide capital for the founding and operating of an ESCO is therefore not entirely without foundation. It could however be overcome by companies proving the profitability of their profession. This requires that existing ESCOs have a good track record. Besides easier access to credit, a larger supply of risk capital would also be beneficial to the development of the profession. As mentioned before, the Commission activities in the field of financial engineering comprise the promotion of venture capital companies in the EC and this is making more risk capital available in Europe.

A guarantee scheme covering the risks linked to the operation of the company - for example a mutual insurance fund - would also facilitate access to operating capital. However, the risk involved in financing energy-saving measures goes beyond normal managerial risks in that it is to a large extent determined by the fluctuation of energy prices. Experience has shown that substantial energy price decreases are possible. These could endanger the profession as a whole as they would not only reduce demand for its services but also jeopardise the success of projects already undertaken. Mutual insurance schemes reducing the normal managerial risk in operating an ESCO would fail under these circumstances. However, a pool of ESCOs operating on the futures market for oil could be one way to reduce the risk connected with the fluctuation of energy prices. Unfortunately, these markets tend to be limited to the short-term.

As you have probably understood, the financial engineering activities of the Commission do not comprise any specific measure for the energy saving sector. However, the approach on the whole in its thrust to revive the spirit of enterprise and even more some of its specific features - such as the measures to increase the amount of venture capital available to industry - could create an environment in which economically sound activities such as energy saving will certainly have their chance to flourish.

ACCELERATING DISCRETE ENERGY EFFICIENCY INVESTMENTS THROUGH THIRD PARTY FINANCING

Dr. D.A. FEE
Principal Administrator, Energy Saving Division,
Commission of the European Communities, Brussels

INTRODUCTION

The Council of Ministers, at their meeting in September of 1986, set new energy objectives for 1995, which included a further improvement in energy efficiency of at least 20 %. This improvement will be effected both by managerial and behavioural changes, and by investments.

Energy efficiency investments can be either integrated investments e.g. new electrical appliances, new cars, new buildings, new burners and boilers, and new industrial processes, or discrete investments i.e. primarily intended to save energy costs. In general integrated investments lead to energy saving as a by-product of some technological innovation.

Although the 1995 energy saving objective of the Council calls for a global energy saving of at least 20 %, the attainment of this objective on a sectoral basis would lead for a requirement for a saving of approximately 52 mtoe in the building and tertiary sector (35 mtoe and 20 mtoe respectively), and a saving of approximately 42 mtoe in the industrial sector. The Commission study entitled 'Towards a European Policy for the Rational Use of Energy in the Building Sector' estimated that the average cost per toe saved each year for existing buildings is 1300 ECUs. Therefore the attainment of the 1995 objective would call for an investment in the order of 67 billion ECU. Similarly the Agence Française pour la Maîtrise de l'Energie has estimated that in the industrial sector an investment of 1050 ECU is required to save 1 toe. Therefore, the total investment required to attain the Community's 1995 objective in the industrial sector will be in the order of 44 billion ECU.

Taking the building and industrial sectors (because of it's complexity the transport sector has been excluded from this analysis) together a combined investment of the order of 100 billion ECU will be required to attain the improvement set out in the 1995 objectives.

One may assume that discrete energy efficiency investments make up only a part of the total investment required, and that measures in managerial and behavioural change, and integrated investment will continue to bare fruit. Nevertheless, the shear scale of the required investment necessitates the development of financial instruments, other than direct State intervention, which will assist in accelerating the discrete investment in energy efficiency.

A Commission communication entitled 'Towards a European Policy for Energy Efficiency in the Industrial Sector' has already examined some of the factors which militate against discrete energy efficiency investments, these include :

- low energy prices ;
- the low priority attached to energy saving investments in decision making processes ;
- lack of knowledge of consumption ;
- financial structure of firms, lack of finance ; and
- the disparity of required rate of return between energy supply and energy savings projects.

For a novel financial mechanism to be successful it must counter all, or most of these factors.

2. THIRD PARTY FINANCING

Several financial mechanisms have been developed to accelerate energy efficiency investments. These include :

- innovative vendor financing, e.g. financial savings guarantees, vendor backed equipment leasing, package financing, and shared saving contracts;
- energy service company financing, e.g. third party financing ;
- energy project financing ; and
- utility financing.

Each type of financing uses different mechanisms, involves various technologies, and can involve more than two participants at the contractual level.

Energy service company financing, or third party financing, has exhibited potential for mobilising the large amounts of private capital required to carry out discrete energy efficiency investments. This mobilisation of private capital is accomplished by an outside company funding an energy saving programme using the cost savings themselves to pay for that investment. Therefore, the energy savings are viewed as a 'stream of income' which can support a business : the business of investing in, and providing performance guarantees for energy conservation, by an energy service company (ESCO).

The concept of the energy service company is, of course, central to the successful operation of the third party financing mechanism. An ESCO must provide a combination of engineering and financial skills. It must be capable of carrying out detailed energy audits, and of suggesting technologies which would be suitable for making planned energy savings. Project finance must be raised, and the flow of funds from the project should be sufficient to repay the provider of the finance, and ensure the profitable operation of the ESCO. In general an ESCO has been defined as a company which 'provides the service of auditing, installation, operations, maintenance and financing on a turnkey basis'. A company which sells equipment but which does not finance or maintain that equipment is not considered to be an energy service company.

The various steps necessary in establishing a third party financing investment are outlined in Annex 1. The ESCO carries out an initial walk through energy audit to establish the level of possible energy savings. A proposal is then made to the facility owner which outlines a programme for accomplishing these energy savings. A contract is negotiated, and an energy baseline or average consumption pattern is developed. The ESCO then carries out a detailed energy audit, and installs equipment aimed at accomplishing identified energy savings. The facility owner and the ESCO share the financial benefit from energy savings made during the term of the contract. Adjustments may be made to the terms of the contract anytime during the life of the contract. When the contract expires, the facility may renew the contract at an adjusted share of savings, he may become the outright owner of the equipment, or he may have an option to purchase the equipment at a price decreed by the contract.

2.1. The Advantages and Disadvantages of Third Party Financing

Third party financing has the following advantages ;

- the facility owner does not have to raise up-front cash to finance conservation measures ;
- the third party assumes all the risk that energy savings will occur ;
- the facility owner does not have to determine which equipment is most

appropriate for their facility ;
- the facility owner can still make alternate investments while starting to reap the benefits of energy saving ;
- it is usual for the facility owner to own the equipment at the end of the contract or arrangements can be made to secure equipment ownership.

The disadvantages of third party financing are :
- third party financing contracts tend to be complex, resulting in a number of facility owners being discouraged from attempting such schemes ;
- the lack of historical energy consumption data can become a limiting factor in the conclusion of a contract ; and
- the economics of third party financing appear more suitable for large programmes than for small ones.

3. THE EXPERIENCE TO DATE WITH THIRD PARTY FINANCE

Third party finance was originally developed in North America, and therefore much of the operational experience has come from there.

The market for third party financing in the United States has been developed over the period 1981-1986. In 1980 there were about 20 companies offering 'energy services' in the United States. Energy saving investments made through these companies resulted in about $ 1m being invested. By 1984, the number of companies had grown to 150, and annual investment stood at some $ 350m.

One of the factors which has assisted the growth of the 'energy services' market in the U.S. has been the active role played by government-Federal, State, and local. The active participation of government institutions has led to a situation where by 1985 the public sector accounted for 50 % of all third party financing compared to 20 % in 1983. At Federal level, the government has, through it's various Departments promoted the use of third party financing in making energy saving investments in government buildings. The Federal Energy Management Programme has set up a clearing house on third party financing, in order to assist government building managers to avail of the technique.

At the State level, programmes have been developed to guide building managers on the utilisation of third party financing to reduce energy consumption in State run buildings.

At local level, many of the County administrations have supported schemes aimed at demonstrating the efficacy of third party finance for energy saving investments in public buildings and in individual homes.

Since the inception of the third party financing technique in the U.S., many disparate organisations have commenced operations as providers of third party financing services. They include ; engineering consultants, equipment manufacturers, subsidiaries of gas and electric utilities, and in some cases by local governments themselves.

In Europe, the concept of third party financing has been much slower to develop. A study carried out for the Commission in 1986 found that the technique is virtually non existent. In 1985, the two large and several small ESCOs collectively invested about 16 million ECU in energy saving projects. Only four countries, the U.K., Belgium, Spain, and Luxembourg had any direct involvement in third party financing while France and Italy had experience with complimentary financing techniques. The 1985 investment figure can be contrasted with the EUR-12 potential of 111.6 billion ECU for energy saving investments.

4. THE BARRIERS TO THIRD PARTY FINANCING

Several factors have been influential in restricting the more widespread utilisation of third party financing in the EUR-12. Among the major factors are :

- lack of finance. ESCOs in both the U.S. and Europe have tended to draw their finance from a larger parent company. In some cases venture capital has been used to support the creation of an ESCO but thusfar the traditional suppliers of capital, the financial institutions, have been unwilling to support the operations of ESCOs. The reasons for this are twofold. Firstly these institutions are ignorant of the operation of the third party financing mechanism. Secondly, while financial institutions have a considerable level of experience in the provision of energy supply project finance they have, as yet, little or no experience in the field of energy saving programmes. However, the risks associated with energy savings, e.g. changes in oil price, are similar to supply projects. Therefore there is no reason why financial institutions should not become conversent with energy saving project risk after some exposure. One level of risk which is difficult to quantify is the technical capability of the ESCO. There is, therefore, a confidence gap between the ESCOs and the financial institutions which can only be filled by working successfully together ;
- lack of knowledge of the technique. To date the limited application of third party financing in EUR-12 has been caused by the mechanism not been widely understood, or even known ;
- complexity of contract. Many third party financing contracts appear complicated to those wishing to make energy saving investments. This apparent complexity has turned many potential investors away from utilization of the mechanism ;
- there are some administrative problems which have restricted the application of third party financing techniques. There has been the example in one Member State, where a decision by the Treasury Department that third party financing contracts entered into by local authorities would be considered as expenditures by the authority for that year, and would therefore form part of the authority's budget. This ruling effectively blocked any third party financing investment by the Member State's local authorities.

A NEW SOURCE OF FINANCE FOR INVESTMENTS IN ENERGY SAVINGS

J. JUNKER
Bayerische Landesbank Girozentrale München

Energy is an essential precondition for the survival of any human being or country, indeed for the entire world today and in the future.
Energy affects our lives in many different ways.

- As private individuals we need heating for our houses, petrol for our cars, electricity for lighting, the washing machine and the model railway, to mention but a few.

- The entrepreneurs among us need the help of energy to power our machines, data processing equipment and transport fleet on the one hand, while it represents a far from negligible cost factor on the other.

- For our national political economies energy is often the key to many problems. Growth of GDP depends to a considerable extent on the conditions of availability of energy. Energy prices are a major factor in the balances of trade and payments of many countries.

- There are a limited number of non-renewable energy sources on "spaceship earth"; economical and rational use of those resources, also out of environmental considerations, should therefore be our guiding principle.

The objective at saving energy was successfully achieved after 1973 and 1979; the 20 % reduction in energy use, called for by the Council of Ministers of the European Community, was achieved by 1985. A further reduction, also a Community objective of 20 % by 1995 is proving more difficult, as little more can be achieved with the relatively simple means applied so far. The sustained interest on the part of the public authorities in saving more energy is clearly reflected in the many different forms of available support (subsidies, tax concessions and long-term credit not only at Community and Member State level, but, within Germany also at Federal end land level.

In addition, new, unconventional methods are appearing on the scene. One of these is TPF, or third party financing, a system developed in the United States, for investing in energy-saving programmes based on the principle whereby an outsider puts up the money and the return on his investment is financed from the energy savings thus achieved.

Introducing this concept in the Community - particularly in the Federal Republic - has run into a number of difficulties :

- the prime objective of investing in industry, for energy saving purposes or otherwise, is to make the investment pay. Investment in energy-saving measures, however, often take more than the (in industry) usual five years to mature. Consequently such investments are low on the list of priorities of many firms and tend not to be made;

- investment in housing, however, is also made with a view to making a profit. Yet here considerably longer amortization periods are involved. Since a great deal has been done in recent years. e.g. in terms of insulation - not least due to public support measures - the scope for more energy-saving measures in this sector is getting smaller.

  Investment in energy-saving measures in terms of heating usually meet with scepticism on the part of the consumer. The level of motivation goes up and down with the heating bills, i.e. energy prices. Thus, the attempt by a well-known German electricity supply company, to sell heat pumps through a subsidiary, were a resouding failure in spite of considerable financial support;

- in the Federal Republic relatively few investors are prepared to go into "third party financing" deals, since the risk factor is hard to assess.

Financial institutions, particularly banks, represent the majority of potential investors.

Banks, however, apply strict criteria - particularly where risk assessment is concerned - as is clear from the conditions of the Federal Credit Control Office.

With regard to risk assessment a distinction can be made between two types of credit.

1. Banks, like the Bayerische Landesbank, often invest in projects such as refuse burning heat and power generating schemes or biogas utilization schemes. in which hydrocarbon bases energy forms such as coal, oil or gas, are replaced by energy derived from the treatment of waste products. The risk in these cases is minimized because of the nature of the borrowers (public bodies) and the contractual commitment to which they are bound.

2. The other group - what one might call traditional financing of energy-saving measures by private enterprise - is more complicated, since the borrowers tend to be small or medium-sized businesses, whose credit worthiness (and therefore credit risk) can often not be accurately assessed for the entire period over which the credit is extended. The liquidity of large firms, on

the other hand, is usually such that investments of this kind can be made without major borrowing requirements.
If banks which feel themselves duty bound for reasons of political economy to support investment in energy-saving schemes by means of specific credit arrangements are to be made to provide such credit also in problem cases, ways must be found of minimizing risk.
In this context talks were held between the VIK, Verband Industrieller Kraftswirtschaft E.L. (Association of German Power Industries) in Bonn, and the Federal Ministry for Industry.
The possibility of Federal guarantees were discussed in detail.
Civil servants pointed out that up to a certain level of investment such guarantees were primarily the responsibility of the Länder. The Federal Ministry will, however, take steps to put this matter on the agenda of the appropriate body dealing with such guarantees at Land level. At the same time the Federal Industry Ministry will accept in principle the possibility of Federal guarantees to underwrite investments aimed at the rational use and saving of energy - including third party financing.

Third party financing of energy-saving schemes is - in addition to existing public support measures - definitely an interesting alternative way of achieving further savings. If, as we have shown earlier, the risk factor for investors can be reduced, and the concept of third party financing becomes more generally accepted. a not inconsiderable contribution can be made to the Community's aim of reducing the use of energy by another 20 % by 1995.

FINANCING INVESTMENT IN ENERGY EFFICIENCY FROM THE MANUFACTURER'S POINT OF VIEW

Phoebus KALYVAS
Motor Oil (HELLAS)

Mr. Chairman, Ladies and Gentlemen,

I had originally intended to contribute to this round table by describing how a large industrial company in Greece successfully financed an investment in energy efficiency.

I subsequently felt that it might be more useful to comment on the difficulties which are sometimes encountered in financing such investments in the industrial sector. Since substantial energy savings can be made, not merely by companies which are major consumers or highly capital-intensive but also, and above all, by small and medium-sized firms with low energy consumption, we should perhaps examine the position of SMEs and the problems facing them.

There are two main reasons why firms fail to invest in energy efficiency :

a) <u>It is not a company priority</u>, i.e :
   - i) energy costs are not always a major component of production costs as a whole ;
   - ii) however encouraging the energy efficiency ratio, the return on the investment is often unsatisfactory in the medium term, particularly where soft or renewable energy is concerned.

b) <u>Simultaneous investment in energy efficiency and company objectives frequently exceeds financial resources</u>
   - i) Most manufacturers have to restrict their investment budgets to projects which are as cost-effective as possible in the light of the company's growth strategy and financial equilibrium.
   - ii) Companies which cannot finance investments in energy efficiency internally depend on external sources such as commercial or institutional banks.

Commercial banks usually require guarantees which are sufficient to cover the commercial risks of the company's activities as a whole (even if the risks involved in the energy project are minimal and the cash flow generated is enough to cover the costs).

Moreover, institutional banks in Europe (specializing in investment) may impose even more stringent conditions such as a guarantee from a major bank. In some countries and in certain circumstances, even a guarantee from one of the largest national banks is not considered adequate.

Private industrial firms therefore often encounter serious obstacles when attempting to finance this type of project. The guarantees required (e.g. mortgages, personal or collateral guarantees, etc.) are difficult to secure and add the cost of the finance.

In addition to direct applications to banks by the companies concerned, there are other forms of financing such as leasing or third party finance. Although these are available to other sectors of the economy such as the building sector, public enterprise or local authorities, they are not always applicable to private industry.

<u>Leasing</u> : the equipment to be installed does not always have a commercial value independent of the rest of the plant and so does not always constitute an adequate guarantee.

Third party financing transfers the problem of covering the company's commercial risks to a third party. If the latter lacks sufficient funds or is unable to use savings for high-risk investment, it has to resort to the banks and comes up against the same difficulties.

Funds for implementing certain projects are, of course, obtainable from the Community directly or via some national governments, but these are usually insufficient and subject to lengthy negotiations.

To conclude, it seems to me that certain obstacles to the financing of investment in energy efficiency in industry, particularly in private enterprise, need to be removed before any improvement can be expected.

In addition to the encouragement of new financing methods such as third party finance, I feel that some kind of Community guarantee to cover specific projects would be a useful aid. This could be proportionate to the energy savings expected from the project and should, of course, be granted only to companies which are normally viable and competitive, and under certain conditions.

A fund consisting of the income from guarantees issued (normal guarantee premiums paid by the recipient companies) could be set up to cover most of the risk incurred by the Community.

DISCUSSION AND CLOSING SPEECH

Discussion

Closing speech by M. DAVIS

DISCUSSION

MORNING SESSION OF OCT. 19
CONTRIBUTIONS of Mr. LE GOFF, LINNHOFF, DAVIS, SCHAEFER, KALIVENZEFF

The question was raised how the concept of exergy could improve a correct and quantitative understanding of the energy saving optimization processes. Also how similar sophisticated technical languages could be accepted in small industries having energy saving problems and lacking of a specialist's staff.

It was argued that in the energy saving processes, the general use of the concept of exergy, a non-conservative, $I-T_2/T_I$ dependent, thermodynamic quantity which is actually "consumed" during the thermal degradation, appears to be more fruitful than the current concept of energy, a T-dependent quantity which is conserved, although thermally degraded. In particular, it was suggested that the "Pinch Point" technique could earn a significant quantitative support if T/H diagrams were subsituted by $(I-T_2/T_I)/H$ diagrams. The answer was that T/H diagram substantially contains adequate information in most cases ; it was however recognized that the exergy concept is particularly useful where chemical energy transformations occur. It was finally assessed that the "Pinch Point" technique is really powerful so that its utilization ought to be widely diffused both in the academic and in the industrial context. As far as the problem of the small industries is concerned, their difficulties in approaching these complicated techniques can be overcome by consulting engineering organizations. The formation of these specialists requests at least two years of intensive training of already experienced engineers.

AFTERNOON SESSION OF OCT. 19
CONTRIBUTIONS of Mr. SCHAEFER, NOLTING, DAVIS, LE GOFF, GLUCKMAN, HERSTEL, REAY, HUYGUES

Some important and general questions arose, concerning in which extent the new type heat exchangers reported by Mr. Reay and Huygues are commercially available. Referring to Mr. Gluckman's paper it was asked how the European Industry will withstand to the Japanese competition in the field of the big absorption heat pumps.

More specific questions concerned how to control the acid condensate emission from low temperature heat recovery exchangers in a brewery (Mr. Nolting paper) and which are the pressure standards in the plastic exchangers (Mr. Huygues paper).

About the commercial availability of the new type heat exchanger it was answered that compact heat exchangers still need some R & D. Catalytic exchangers are to some extent already used in specialized aircraft conditioning applications. Small plastic exchangers with millimetric tubes can be used, without occlusion risks, only with pertinent fluids (e.g. in the pharmaceutical industry).

As far as the Japanese competition is concerned, it was observed that Japan is presently producing big size (350 to 5000 kW) heat pumps, 85 % of which of the absorption type. An important effort is in progress in order to reduce to about 70 kW the limit of economic convenience of the absorption type. In Europe this type is present only in few examples. Different ratios between electricity and heat costs in the various countries displace the limits of convenience, compared with the Japanese situation. It was assessed that the European industry can develop a valid competition, based on important technological sectors, like screw compressors and others, where Europeans are more advanced or as advanced as the Japanese.

On the more specific questions it was argued that daily tube leach-washing allows brewery's condensate neutralization and free discharge in the ambient area.

Pressure standards of rigid capillary-tube-exchangers range from 4 to 6 bar max. at about 100°C. Plate exchangers allow 0,1 bar at 5-10°C temperature difference, which are suitable for the thermal vapour recompression.

FIRST MORNING SESSION OF OCT. 20

CONTRIBUTIONS of Mr. DAVIS, VAN HEEL, HERSTEL, BARENDREGT, SCIARRETTA, LUXTON, KALIVENZEFF, ARENDT, PILAVACHI, BUNGE, LE GOFF

The very important question was stressed, how could the experience gained in the Demonstration Programs sponsored by the Commission, influence the industrial replications of the plants both from the economic point of view (pay back) and from the technical point of view (plant structure).

Other general points of interest were raised on the availability of mathematical methods of optimization for the process as a whole (Sequential simulation technique) ; on the impact of complex plant additions due to the energy saving process on the global plant's reliability and maintenance costs. More specific questions were asked (Mr. Kaizer paper) on the king of R & D requested for the specific optimization of the chemical reactor for the steam cracking and on the type of computer utilized for determining the flow data.

The utilization of the experience of the Demonstration Programs in the industrial replication was judged as a very difficult question. As a matter of fact, the replication's economy strictly depends from plants local specificity. There are basic differences between a demonstration plant and an industrial plant, which must fullfil stringent product quality standards and environmental respect, beyond energy saving goals. These last-ones vary according to energy price-policies and to plant's load factors.

There was evidence, in the various contributions, of almost random scattered answers, according to the choices on the numerical values of the various techno-economic parameters.

Concerning the availability of global optimization (process integration) methods, there was agreement on the fact that modelling the whole plant still presents some difficulties. The positive fallout on the engineers was stressed, who learn more from models than from papers. It was also argued against the magic expectation that computers can give any possible answer.

Concerning the impact of energy saving's plant additions on the whole plant reliability, an acceptable impact was reported from the Italsider (Mr. Sciarretta paper).

To the specific questions to the Mr. Kaizer paper it was answered that variable feedstock may affect the reliability of the model's prediction. It is felt that some R & D could still be required.

The computer used is a HP 9000. Systems of about 4000 equations request about twenty minutes computer time, provided that this is fully available.

Finally, some considerations were presented concerning the difficulties in obtaining financing from the companies management on such computerized methods, especially when results hardly match expectations, due to rapid changes in energy prices-policies and when production programs are strongly modified in the short term.

Also the complex interaction between man and computer was evidenced, whose critical path presently passes through the very slow interface of the keyboard-display arrangement.

SECOND MORNING SESSION OF OCT. 20

CONTRIBUTIONS of Mr. TRIOULLIER, ZORZOLI, GOUBLOMME, COEYTAUX, PAULES, SPRINGELL, DAVIS, GREHIER, CITTADINI, KALIVENZEFF, LE GOFF

The important question was stressed, whether it is possible to introduce quantitative evaluations in the calculations of the benefits induced by the environmental emission control (Reference to Mr. Zorzoli paper).

More specifically referring to computerized process optimization methods (Mr. Coeytaux and Springell papers) some questions were presented on the possibility to introduce electricity fixed costs and contracted power effects ; component's new technology effects ; price reduction effects due to big quantity purchasing ; extra saving effects when the computerized optimization is contemporaneously carried on the process and on the utility side. It was also asked whether the calculation's accuracy of the model in determining flow balances exceeds the instrument readings and also which fraction of the total cost of the model pays the measuring sensors.

Referring again to Mr. Zorzoli's paper, it was asked which is the role of the reciprocating motors in the cogeneration units (CHP) expected in the next decade.

On the first question it was pointed out that legal regulations substantially differ in the various countries, thus producing different constraints on quantitative evaluations. A concrete judgment could probably only be stated on the basis of the owner's availability to close the plant (cost) against environment control investment (benefit).

The group of questions concerning the computerized methods were positively answered. It appeared that the algorithmic potential is easily adequate, the only difficulty consisting in a correct description of the continuous or discontinuous functions representing the various quantities. It was stressed that the objective function to be optimized is, in any case, unique, the remaining functions being considered as constraints. It was also outlined that there are of course errors in the flow balances in spite of the good quality of the instrumentation. Some data were finally offered about callated savings : about 5 % in sugar and paper pulp factories ; 3 to 5 % in utilities. Higher savings are typical of batch processes rather than in continuous processes, where the methods of calculation are similar to those ones of the utilities. When utility and process integration is considered, optimized savings may be higher but costs are also higher.

About the role of reciprocating motors in CHP it was outlined that there are many industrial producers in Europe, also in the smaller countries. Larger utilizations are expected in the lower power, lower temperature range.

Finally, a new type of a possible future utility was presented, namely the utilization of low grade energy, such as waste heat, night electricity etc. for separating dilute solutions (e.g. ammonia from water or glycole from water, or a light hydrocarbon from a heavier one) and to recover it in an up-graded form, after stockage, when mixing again the solution's components. Though the economy of such a process is still generally negative, there is evidence of a good future potential and there are some specific cases where a benefit is already possible. Joint research is on progress in Europe and in Japan.

ROUND TABLE, OCT. 20
CONTRIBUTIONS of Mr. DAVIS, FABER, KALIVAS, JUNKER, FEE, MOERDIJK, MOROVIC, SAKELLARIOS, HADJICHRISTOU, JANDEAU, FEDELE, KLEEFKENS

From the context of the discussion, following themes appeared to be of major interest :

- Is the European Commission's remarkable financial and informative effort adequate to the present development of energy efficiency's technology ?
- How could a wide third party financing effort be provided and which kind of guarantee would be required in order to enhance the energy efficiency program to an industrial scale ?
- Would the techno-financial activity of the ESCO's (Energy Service Companies) be of some help in order to facilitate such enhancement ?

On the first theme, a proper source diversification and an efficient use of the energy is stated as a priority task of the European Community. The Commission receives funds from the member countries and delivers them back informations on the energy efficiency technology. The financial effort was, till now, of about 750×10^6 ECU on about 1500 projects sponsored. Difficulties in project replication were sometimes experienced, due to strong economic dependence from local situations.

The Community cannot be considered as an unexhaustible source of credit : it has limited funds, its 1988 budget is not yet approved. The Commission priority as a supporting institution must be slowed down. Moreover, the member countries are reluctant to increase the risks assumed by the Commission in form of guarantee, insurance primes etc.

New third party financing is needed in order to enhance the industrial development of energy saving, where a big amount of money is requested.

It is felt that adequate sensibilization, information and diffusion of the concept itself, and of the potential benefit of energy efficiency process, are of basic importance, even more than financial problems.

Presently, the European Commission compels its contractors to diffuse and publish the results of the successful programs. Moreover, a data base, named SESAME is available, which contains a summary of the Project's information and progress state, and which can be used for information retrieval. It is recognized that this activity could be further improved.

On the second theme it was assessed that various significative distinctions can be envisaged, when speaking about energy saving industrial projects, requesting financial support. For instance :

- Highly innovative technology versus low innovative or conventional technology. The risk involved in financing such programs ought to be considered as moderate, since it is recognized that energy saving normally involves medium innovative or conventional technology.
- Space heating versus remaining types of energy saving. Space heating represents an important fraction of the global saving potential. An

easier account of the benefits and risks is involved in space heating programs, thus favorizing a possible third party financement. Typical cases have been illustrated.

- Big companies versus medium-small size companies realizing the energy saving investment. Big companies, due to their larger financial liquidity some time don't need third party financement. But if they do, then they implicitly offer better confidence and warrant of societary stability.

From the extensive discussion, it became evident that banks are reluctant to risk money in the energy saving business. The risk of fuel price instability, which is involved in the more general case of financement of energetic programs, seems to be more acceptable than the risk of energy saving, which depends from price differences.

Among the distinctions presented above, the banks seem to be most sensible to the sharing of big versus small companies. It is a psychological factor : a big oil company can loan billions DM. But how to make confidence in a small, even high technology firm, whose activity is not well known, whose assets only cover a fraction of the loan, and whose existence is not sure even in the immediate future ?

The problem of guarantees is then imperative. Here again the case of the small firms and of energy saving minor interventions was stressed, due to the difficulty of identifying the object of the guarantee. This can be put on a piece of equipment but not on a technical amelioration.

Various possible guarantee schemes have been considered and proposed: from National Governments, from local Authorities, from Bank Consortia, from the European Commission, even limited to the replication of successful programs. Also insurance forms have been prospected, as well as interventions of the BEI. The scheme of a mutual guarantee, although not better detailed, appeared finally as the most promising, to the audience.

On the third theme it was recognized that the role of the ESCO's (Energy Service Companies) presents different characteristics in the USA and in Europe.

In the USA the ESCO's are more widely diffused since 1981 and now there are more than 250 companies. Most of them are utilities or utilities-dependant firms. They have juridic form in order to enter the field ; they own large experience in auditing and expertizes, and have access to the available public funds. The aim of the utilities in managing their ESCO's seems to be a real reduction of the energy consumption in order to avoid investment increments in their networks.

In Europe this aim seems not to exist ; in the case, possibly to increase energy selling. The European ESCO's are presently about ten.

It is felt that the presence of the ESCO's could facilitate a mechanism of mutual guarantee. But the opinion was also expressed that, on the contrary, the presence of a medium size techno-financial intermediate could increase the reluctancy of the banks to loan money. Higher confidence in the ESCO's could derive from the participation in the ESCO's of the Public Institutions or possibly of the European Commission, or of private Bank Consortia.

Some positive experience in bank financing to ESCO's was however reported even without guarantee, but with very short payback times (Spain) in the field of Hotel energy saving. Strong interest to ESCO's and favourable market conditions for energy efficiency programs are also reported from the Netherlands. Also the doubt was expressed that priority could be given by the Commission, to ESCO's programs operating in bigger bulkenergy saving companies, rather than in smaller but percentually higher energy saving companies. It was recognized that this risk sometimes exists.

Closing Speech by Dr Michael DAVIS

Ladies and Gentlemen,

I have now the task to try to sum up my conclusions of this Confererence on Energy Efficiency in Industry. On this day when frantic selling of shares on the New York Stock Exchange has been reported; and investors must be preoccupied with issues other than investing in the more efficient use of energy in industry, let me make a few observations.

Recent years have seen far too much in share price and currency exchange manipulation to sustain paper profits. Not enough attention has often been paid to investing in the real solid worth of productive industry. Industry is, after all, the motor of our economy.

Among the investments by industry to increase productivity, to introduce new products and to increase sales, investment in the more efficient use of energy is, as Signore Briganti remarked, in an opening address (which he made on behalf of Prof. Colombo, President of ENEA): "energy efficiency is increasingly associated with innovations, but it goes further by increasing the quality of products; is favourable for the environment and helps industry to better accomodate to changing demand".

Today in closing this Conference, I believe those of you who have participated in this 2-days event in Berlin will have obtained a better understanding, and I hope an inspiration, to work for the enhancement of real worth of our industry which intelligent investment in energy efficiency represents.

What then are the inferences to be drawn from this conference ?
I believe they are the following :

- Firstly, at a time when potential instabilities in the world can suddenly confront us with an unforeseen and rapidly changing economic environment, I cannot stress too much the imperative of having a sound energy policy: a policy, as my Director General, Mr Maniatopoulos said, diversifies our sources of energy and ensures that all the energy we use, is used efficiently. The European Community wants an energy efficiency improvement of 20 % by 1995.

- Secondly, our industry must strive to keep in the forefront of innovation, so as to be competitive in the market. Innovation in energy efficiency is one of the important factors in this purpose. Several presentations to this conference have shown that europeans are inventive and innovative in a whole variety of practical energy efficiency applications. But the exhileration of getting such economic innovations applied successfully, must not remain the prerogative of just a select few. It must permiate and penetrate all of our industries. Let us learn from the example of those cases presented at this conference, which demonstrated that such innovation is challenging, achievable and rewarding.

- Thirdly, this conference has shown that the complexity of industrial process plant optimization has now got a remarkably sophisticated tool of analysis which goes further than the term "Process integration". This is typified by the "Pinch Technology" which Prof. Bodo Linnhoff so eloquently explained to us. It rests on the basis of what is virtually new thermodynamic principle, which he discovered, but has now been endowed with analytical and computer aided methods to apply it to both existing and new plants. Its potential has been demonstrated to us with several practical examples. But European industry must grasp what this means for them. Sectors of large American industries have already recognized the value of this European discovery and are applying it profitably. Let European industry open its eyes, be professional and get to work massively at this exciting technology with no more delay.

- Fourthly, many desirable and economically justifiable new investments in energy efficiency are just not being made, particularly in small and medium sized industries. This is quite understandable given more pressing priorities, lack of suitably experienced in-house staff, and saturation of lines of credit available for borrowed money. The roundtable, which I have just had the honour to chair, shows that there is a feasible and responsible way of overcoming this obstacle. It is possible to achieve highly desirable energy efficiency investment by the operation of "third party financing" notably through the operation ESCO's - Energy Service Companies. There is also great scope for gas electricity utilities to engage in these activities.

  Promising though third party financing may be, it must not supplant other methods such as those provided by the European Investment Bank, or by the Commission for example under the Valoren scheme to encourage investment in the less developed regions of the Community.

Ladies and Gentlemen, I hope that you will find my remarks a fair conclusion to this conference and that the conference has been helpful to you. I am particularly grateful to all those who made it possible: the Chairmen of sessions, the speakers, the Berlin conference organizer InnoTec and, of course, my collaborators, notably Dr. Zito and Mr. Sirchis who, though not very evident in the actual proceedings, as you can well imagine, have had months of preparation. We have also had valuable help from the Commission office here in Berlin, headed by Mr. Jaedtke. Lastly it gives me pleasure to thank our hosts: the Berlin Senate, Senator Turner and Staatssekretät Prof. Beitz for their words of welcome and the enjoyable reception they gave us last evening. It has been a splendid occasion to help celebrate Berlin's 750th anniversary, a city which has undergone many vicessitudes over the years. Let us express the wish that the remaining 250 years to attain their millenium will be accomplished in peace and prosperity!

ZUSAMMENFASSUNGEN IN DEUTSCHER SPRACHE

Einführung

Prozeßintegration

Neue Techniken zur Rückgewinnung der Niedertemperaturwärme

Industrieanlagen - Prozeßüberwachung und Optimierung

Energiemanagement in Energieversorgungssystemen

Podiumsdiskussion über Finanzierungsquellen für Investitionen zur Verbesserung der effizienten Energienutzung

GRUNDLEGENDE MASSNAHMEN DER ENERGIEEINSPARUNG

Prof.Dr.-Ing. H. SCHÄFER, Lehrstuhl für Energiewirtschaft und Kraftwerkstechnik der TU München

Im Gesamtbereich der Energieanwendung und so auch beim Energieeinsatz in der industriellen Produktion lassen sich grundsätzlich 5 Wege für eine rationellere Energienutzung aufzeigen, nämlich :

- vermeiden unnötigen Verbrauchs
- senken des spezifischen Nutzenergiebedarfs
- verbessern der Nutzungsgrade von Maschinen und Anlagen
- Energierückgewinnung, wo dies technisch und wirtschaftlich sinnvoll erscheint
- verstärkte Nutzung regenerativer Energiequellen.

Eine unabdingbare Voraussetzung für alle Planungen und Maßnahmen zur rationelleren Nutzung von Energie ist eine Analyse des energetischen Status quo. Derartige Analysen müssen,wenn sie eine geeignete Basis sein sollen, sich auf messtechnische Untersuchungen abstützen gleichgültig, ob es sich um einzelne Anlagen, Maschinen oder ganze Betriebe handelt.

In vielen Fällen können neue Techniken dazu dienen, Energie rationeller einzusetzen, oft genügt es jedoch, bestehendes wissen und eingeführte Techniken richtig zu koordinieren.

Energieeinsparmaßnahmen können energietechnischer Natur sein und werden sich im allgemeinen als Zusammensetzung dieser Faktoren darstellen, um effektive Ergebnisse zu erzielen, ist deshalb eine enge Zusammenarbeit der einzelnen Bereiche erforderlich. Angesichts der oft sehr unterschiedlichen Relationen von Aufwand und Nutzen bei einzelnen Maßnahmen ist in jedem Einzelfall sorgfältig zu prüfen, welcher Weg zu rationeller Energienutzung hier der optimale ist, besondere Beachtung ist dabei der Tatsache zu schenken, daß mehrere Maßnahmen, die in ein- und demselben Bereich ergriffen werden, einander im Nutzeffekt stark beeinflussen können, darüber hinaus müssen auch komplexe Wechselwirkungen zwischen Änderungen der Produktion oder Produktivität und Änderungen im Energiebedarf beachtet werden. Man muß sich klar werden, daß in vielen Fällen Energieeinsatz durch Kapital substituiert wird und daß dabei die eingesetzten Techniken komplexer und die Wirkungsprinzipien immer weniger durchschaubar werden. Die Bedienung und Wartung verlangen deshalb zunehmend geschulte und hochqualifizierte Kräfte und werden zu einem noch bedeutenderen Kostenfaktor als bisher.

ENERGIEEINSPARUNG BEI DER FERTIGUNG VON KURBELWELLEN EIN BEISPIEL FÜR INTEGRIERTE ANALYSE AUF DER GRUNDLAGE DETAILLIERTER MESSUNGEN

Dr.Ing. M. RUDOLPH, Lehrstuhl für Energiewirtschaft und Kraftwerkstechnik, T U München

Um die möglichen Energieeinsparungen in konkreten Fällen zu quantifizieren, bedarf es einer messtechnischen Erfassung relevanter Daten und, darauf aufbauend, einer Analyse des energetischen Betriebsverhaltens der betreffenden Anlage unter den Bedingungen des praktischen Einsatzes.
Es wird über eine Untersuchung zweier Fertigungslinien zur Bearbeitung von Kurbelwellen berichtet, die sich zum einen im Grad ihrer Verkettung, zum anderen im Material und Herstellungsverfahren der eingesetzten Rohteile unterscheiden.
Die Gesamtwerte des Energieverbrauchs je bearbeiteter Kurbelwelle unterscheidet sich für die beiden Fertigungslinien um mehr als 20 %.
Die Ursachen hierfür sind komplex. Von wesentlichem Einfluß sind die Unterschiede in den zu zerspanenden Materialmengen, in der Fertigungskapazität der beiden Linien, in den Relationen zwischen Haupt- und Nebenzeiten, sowie im Verbrauch der Verkettungseinrichtungen.
In einigen Bearbeitungsstufen werden bei beiden Fertigungslinien auch unterschiedliche Bearbeitungsverfahren für ein und dieselbe Bearbeitungsaufgabe angewendet.

Durch Ausschöpfung aller Möglichkeiten der Energieeinsparung wie Abschaltung von leerlaufenden Antrieben, optimal dimensionierte Antriebsmotoren, verlustarme Drehzahlstellung und wahl energetisch günstiger Bearbeitungsverfahren, kann der Energieverbrauch für die Bearbeitung der Kurbelwellen nennenswert verringert werden. Allerdings müssen auch die praktischen Probleme des Aufwandes solcher Maßnahmen und der damit verbundenen Kosten berücksichtigt werden. Auch sollten für die Beurteilung von Möglichkeiten der Energieeinsparung die eventuellen Auswirkungen auf den kumulierten Energieverbrauch, also unter Einschluß vorgelagerter Produktionsstufen, mit in die Betrachtung einbezogen werden.

PROZESSINTEGRATION UNTER VERWENDUNG DER PINCH-TECHNOLOGIE

Linnhoff & Eastwood
University of Manchester
Institute of Science and Technology, Manchester

In den letzten fünf Jahren hat sich gezeigt, daß die Prozeßintregration signifikante Verbesserungen bei der Auslegung von Prozeßanlagen ermöglicht. Dabei wurde nachgewiesen, daß die Integration zu Anlagen führt, die sowohl billiger zu bauen als auch billiger und einfacher zu betreiben sind.

Das Instrument, das diesen Fortschritt bei der Auslegung ermöglichte, ist die Pinch-Technologie. Sie basiert auf der grundlegenden Thermodynamik, ist aber einfach zu verstehen und zu nutzen.

Wir beschreiben in diesem Papier kurz die wissenschaftlichen Grundlage der Pinch-Technologie und zeigen dann auf, wie diese angewandt wird.

Hierzu legen wir zwei Fallstudien vor, die den Nutzen der Prozeßintegration im Zusammenhang mit der Wärmerückgewinnung eindeutig aufzeigen. Die erste Studie stellt die Tatsache heraus, daß die Wärmerückgewinnung im Rahmen der Gesamtanlage (und möglichst des Gesamtwerkes) zu betrachten ist, wenn die richtigen Einsparungen gemacht werden sollen. Energieeinsparungen, die für sich betrachtet einleuchten und gerechtfertigt erscheinen, können verpaßte Gelegenheiten zur Folge haben und sogar den Anwendungsbereich von Energieeinsparungen anderwärts verringern.

Die zweite Fallstudie demonstriert, wie die dank der Pinch-Technologie gewonnenen Erkenntnisse dazu genutzt werden können, einen Prozeß so zu verändern, daß Energieeinsparungen gemacht werden.

Schließlich wird durch Verweis auf tatsächliche Studien, die in den letzten fünf Jahren durchgeführt wurden, auf typische Einsparungen hingewiesen, die dank der Anwendung dieser Technologie sowohl bei Auslegungen neuer Vorhaben als auch bei Nachrüstungen möglich sind.

PROZESSINTEGRATION IN EINER BENZOLRAFFINERIE

R.L. BARDSLEY
Staveley Chemicals Ltd, Chesterfield

Staveley Chemicals betreibt 6 Chemieanlagen mit kontinuierlichem Prozeß in der Nähe von Chesterfield, Derbyshire.

Die BTC-Anlage war ein Brennpunkt für Untersuchungen über Energieeinsparungen. Produkte dieser Anlage sind die beim Raffinieren von Rohbenzol in einem katalytischen Hochtemperatur-Verfahren anfallenden Produkte Benzol, Tolöl und Cyclohexan.

Eine 6 Monate dauernde Studie über die Anwendung der Pinch-Technologie, bei der grundlegende Energieverbrauch mit 1 000 Einheiten angesetzt wurde, ergab folgendes Bild :

	BTC-Anlage	Litol-Anlage	Wärmerückgewinnung aus Rauchgasen
Derzeitige Nutzung	1 000	543	543
Mindestenergieziel	860	421	299
Praktisches Energieziel			323

Eine praktische Verringerung des Energieverbrauchs um 40 % schien - einschließlich der Wärmerückgewinnung aus Rauchgasen - erreichbar.

Die Kosten und Gewinne wurden für verschiedene ingenieurtechnische Optionen veranschlagt. Darüber hinaus wurde eine Übersicht über die Energieverteilung für den gesamten Standort erstellt, die Änderungen des Dampfsystems sowie die Installierung von zwei Turbinen empfahl.

	Kapital	Amortisationszeit
Projekt 2		
6 Wärmetauscher : Litol-Anlage	mehr als UKL 100 000	Etwa 1 Jahr
Projekt 3		
Wärme-Kraft-Kopplung : 2 Turbinen	sehr viel mehr als Projekt 2	Weniger als 4 Jahre

Das Unternehmen erwartet, durch die Ausführung dieser Projekte UKL 750 000 pro Jahr einzusparen.

Die Pinch-Technologie erwies sich als entscheidend für die Neuauslegung eines komplexen Wärmeaustauschnetzes im Hinblick auf einen minimalen Energieverbrauch. Die Komplexität wurde auf einige klare Optionen reduziert, deren Kosten und Nutzen beurteilt werden konnten.

PROZESSINTEGRATION IN EINER BRAUEREI

R. MARSH
Tetley Walker Ltd., Warrington

Dieser Bericht zeigt, wie die vom britischen Büro für effiziente Energienutzung geförderten Energieberater Linnhoff March Ltd. die Pinch-Technologie auf die Verringerung des Energieverbrauchs und der damit zusammenhängenden Kosten einer typischen britischen Brauerei anwandten.

Nach einer einleitenden Beschreibung des Unternehmens und seiner Produkte beschreibt das Papier, wie die Brauerei, die ihre Energie schon jetzt sehr effizient nutzt, in der Lage war, den Wirkungsgrad ihrer Energienutzung im Anschluß an die Umsetzung einer Pinch-Technologie-Studie von Linnhoff March Ltd. weiter zu steigern.

Das Papier stellt die praktischen Anwendungen von Energieeinsparungsmethoden, die dazu genutzt werden, Dampf, Erdgas und Heizöl einzusparen, mit Hilfe von Diapositiven im einzelnen dar. Es zeigt, wie drei mögliche Pfade für Maßnahmen identifiziert wurden und vor allem genau was die praktischen Energieziele jedes Pfades für die gesamte Brauerei waren.

Der Bericht schließt mit einer Zusammenfassung der Lehren, die aus der Anwendung der Pinch-Technologie für das Brau verfahren gezogen werden können und wie die Pinch-Technologie auch für andere industrielle Prozesse mit hohen Energiekosten eingesetzt werden kann.

WÄRMEPUMPE UND MECHANISCHE WIEDERVERDICHTUNG VON DAMPF FÜR INDUSTRIELLE ZWECKE

A. GLUCKMAN
March Consulting Group, Windsor

In den letzten 10 Jahren wurden in Europa mehrere hundert industrielle Wärmepumpensysteme installiert.

Die Leistung dieser Systeme war unterschiedlich. Einige Anlagen waren sowohl in praktischer als auch in wirtschaftlicher Hinsicht sehr erfolgreich, andere waren infolge des geringen Ertrags des investierten Kapitals und Zuverlässigkeitsproblemen weniger zufriedenstellend.

Aus den bestehenden Anlagen können viele nützliche Lehren gezogen werden.

Das Papier gibt einen kurzen Überblick über vorhandene industrielle Wärmepumpen. Aus diesen Daten werden Leitlinien für eine gute Auslegung von Wärmepumpen abgeleitet. Das Papier stellt vor allem einige Aspekte des System- und Komponentenentwurfs heraus, die in den ersten Phasen eines Wärmepumpenprojekts stets berücksichtigt werden sollten. Eine Erörterung der Märkte und Einsatzmöglichkeiten von Wärmepumpen in Europa wird vorgelegt.

AUSWIRKUNGEN DER NEUEN TECHNOLOGIEN AUF DIE KONZEPTION KÜNFTIGER WÄRMETAUSCHER

D. REAY
David Reay & Associates, Whitley Bay

Die Technologie der Wärmerückgewinnung stand in den meisten größeren industriellen Energieeinsparungsprogrammen an hervorragender Stelle. Eine signifikante Einbeziehung von Wärmeaustauschanlagen in Prozesse wurde erreicht. Dennoch gibt es noch immer technische und wirtschaftliche Schranken, die einer umfassenderen Einbeziehung von Wärmetauschern im Wege stehen. Probleme im Zusammenhang mit organischen Oberflächenverschmutzungen und der Korrosion bleiben bestehen, und die Versuchung, ausgeklügeltere Verfahren für die Rückgewinnung von Energie, wie ORC-Maschinen einzuführen, hat zu einigen übertrieben ehrgeizigen Anlagen mit zweifelhaftem wirtschaftlichen Nutzen geführt.

Die Aussichten einer kostenwirksamen Niedertemperatur-Wärmerückgewinnung verbessern sich dank einer Kombination von Entwicklungen, zu denen auch der Einsatz neuer Werkstoffe und neuer Wärmetauscherkonzepte gehört. Periphere Hilfsmittel - vor allem verbesserte Auslegungsverfahren und der Einsatz von Techniken der künstlichen Intelligenz - können den Nutzern behilflich sein, dem Stand der Technik entsprechende Anlagen auszuwählen.

Die Prozessintensivierung, eine Phase, die in der Regel mit kompakten Chemieanlagen assoziert wird, hat künftig eine wichtige Rolle zu spielen, da sie eine Verringerung der Größe und Kosten von Wärmetauschern ermöglicht. Dies ist eines der Konzepte, die in dem Papier beschrieben werden, das sich mit den Problemen der Niedertemperatur-Wärmerückgewinnung befaßt und die möglichen Auswirkungen neuer Technologien auf die Auslegung künftiger Wärmetauscher erörtert.

ENERGIENUTZUNG DURCH MECHANISCHE WIEDERVERDICHTUNG VON KOHLENWASSERSTOFFDÄMPFEN

J.P. Livernet
Rhône Poulenc, Chalampé

Zweck dieser Anlage ist es, den an der Spitze einer Destillationskolonne freigesetzten organischen Dampf zurückzugewinnen, ihn auf einen hinreichenden Druck zu verdichten und ihn statt des 6-bar Dampfes aus dem Anlagennetz in den Destillationsgefäßen derselben Kolonne auszukondensieren.

Die Anlage ist in eine Cyklohexanol-1-Produktionseinheit integriert, die seit 1972 im Werk CHALAMPE (Frankreich) von RHONE POULENC betrieben wird.

Die Einheit besteht im wesentlichen aus einem Verdichter, Wärmetauschern und Rohrleitungen. Angesichts der Korrosionsgefahr ist die gesamte Anlage, einschließlich des Verdichters, in Rostfreistahl ausgeführt.

Eine Energiebilanz der Kolonne ergibt eine Einsparung von 32 t Dampf/h sowie einen Elektrizitätsverbrauch des Verdichters von 3 400 kW/h.

Die Auslegungsstudien wurden von der Zentralabteilung für Ingenieurwesen von RHONE POULENC durchgeführt. Ein detailliertes, maßstabsgetreues Modell (1/33) wurde für Detailuntersuchungen und die Ausbildung von Betriebspersonal eingesetzt.

Die Gesamtkosten bis Januar 1985 beliefen sich auf 41 Mio FF. Das Projekt hat in Anbetracht der innovatorischen Aspekte Zuschüsse von der Kommission der Europäischen Gemeinschaften und von der französischen Agentur für Energieeinsparungen erhalten.

Der Betrieb ist trotz einer infolge der Natur der (feuergefährlichen und korrosiven) Fluide erforderlichen hohen Investition erfolgreich.

WÄRMETAUSCHER AUS KUNSTSTOFFEN

J. HUYGHE
G.R.E.Th., C.E.N., Grenoble

Obwohl die Wärmerückgewinnung bei niedrigen Temperaturen im Vergleich zu hohen Temperaturen oft sehr einfach zu sein scheint, ist sie mit einigen Schwierigkeiten verbunden. Es sind dies die Kosten der Wärmeaustauschfläche sowie die Korrosion.

Angesichts der Entwicklung von Kunststoffen erlaubt es das niedrige Temperaturniveau entsprechender industrieller Anwendungen (unter 150 °C) nunmehr, diese Stoffe für die Realisierung von Wärmetauschern einzusetzen, die eine Lösung der vorstehend genannten Probleme ermöglichen können.

In den letzten Jahren wurden daher in Europa verschiedene Arten von Wärmetauschern aus Kunststoffen entwickelt. Dank ihrer spezifischen Auslegung verringern einige Wärmetauscher den Nachteil der schlechten Wärmeleitfähigkeit des Materials auf ein Mindestmaß. Sie weisen im Vergleich zu Wärmetauschern aus Metall folgende Vorteile auf : häufig geringere Kosten, leichtes Gewicht, bessere Beständigkeit gegen korrosive Medien und mitunter geringere Oberflächenverschmutzung infolge der geringeren Adhesion der Verschmutzungsprodukte auf Kunststoffen.

Der Bericht gibt einen Überblick über die verschiedenen Arten von Wärmetauschern aus Kunststoffen, die in den vergangenen Jahren entwickelt wurden. Die meisten von ihnen sind inzwischen für Betriebstemperaturen bis zu 100 oder 150°C auf dem Markt eingeführt worden.

BRÜDENVERDICHTUNG IN EINER BRAUEREI

I.V. NOLTING
M A N TECHNOLOGIE GMBH, München

1. Einleitung

Der erste Erdgasmotor-betriebene Schraubenverdichter für die Brüdenverdichtung wurde in der Privatbrauerei Dortmunder Kronen in Deutschland installiert. Besondere Anforderungen für die Modernisierung des Sudhauses waren hohe Reduzierung der Geruchsemissionen und die Energieeinsparung.

2. Beschreibung der Brüdenverdichtungsanlage

Ein Prozessgasschraubenverdichter der mit einem Erdgasmotor angetrieben wird, wird zur Verdichtung der Brüden eingesetzt. Die Motor-Kompressor-Einheit wurde auf einer Bühne oberhalb der zwei Würzepfannen installiert und mit einer Schallschutzhaube versehen. Das komplette Aggregat ist schwingungsisoliert aufgestellt, so daß eine Körperschallübertragung im Gebäude nicht hörbar ist. Hinter den zwei Sudpfannen ist der zusätzliche Würzekocher - Thermostar - aufgehängt.

3. Funktion des Brüdenverdichtersystems

Das Brüdenverdichtersystem fährt alternativ auf Würzepfanne 1 oder 2. Die Verdampfungsrate wird über die Drehzahl des Gasmotors eingestellt. Der aufsteigende Dampf aus der Würzepfanne mit 1 Bar 100°C wird vom Schraubenkompressor angesaugt und auf 1,3 - 1,6 Bar verdichtet. Dieser Dampf kondensiert im Thermostar und überträgt die Wärme an die Würze, welche dann in die Würzepfanne zurückgeleitet wird.

4. Betriebserfahrung

Die Reduzierungen der Emissionen und die Energieeinsparung wurden im Dauerbetrieb bestätigt. Die Anlage ist inzwischen mehr als 5 000BH gefahren.

5. Schlußwort

Die Zielsetzungen der Sudhausmodernisierung wurden mit dem Gas-Motor-Brüdenverdichter voll erreicht.
In der Zwischenzeit sind in Europa acht weitere Anlagen, davon allein sechs in Deutschland, installiert worden.

NUTZUNG DER RESTWÄRME BEI DER SOLEVERDAMPFUNG

P.E. BUNGE
Akzo Zout Chemie, Hengelo

Mehrstufen - bzw. Mehrkörperverdampfer werden dazu genutzt, um "Vakummsalz" aus Sole herzustellen. Die Sole wird im ersten Verdampfer durch Durchführung von Niederdruckdampf durch einen angeschlossenen Wärmetauscher erwärmt. Der Dampf der siedenden Sole wird bei diesem ersten Körper dazu benutzt, die Sole des nächsten Körpers aufzuwärmen, der bei einem niedrigeren Druck betrieben wird. Dieser Prozeß setzt sich durch die gesamte Batterie der Verdampfer fort, wobei der Dampf des letzten Körpers mit Kühlwasser auskondensiert.

Vier von sechs Körpern werden in der Regel für die Salzgewinnung eingesetzt.

Wenn die Energiepreise steigen, kann es sich als wirtschaftlich erweisen, vor dem vorhandenen ersten Verdampfer einen zusätzlichen Verdampfer zu installieren.

Wenn die Salzgewinnungsanlage jedoch an ein Kombikraftwerk angeschlossen ist, geht der Vorteil eines geringeren Dampfverbrauchs weitgehend verloren, da der zusätzliche Evaporar Damps mit einem höheren Druck benötigt und dementsprechend weniger Elektrizität erzeugt wird. Darüber hinaus muß die Temperatur des ersten Körpers begrenzt werden, um Korrosionserscheinungen zu vermeiden.

Der Dampf des letzten Körpers hat eine Kondensattemperatur von nur 43°C, enthält aber infolge der verfügbaren großen Menge auch eine beträchtliche Menge Wärme. Diese Wärme wird infolge ihrer niedrigen Temperatur im allgemeinen als Abwärme angesehen. Unsere Aufgabe bestand daher darin, einen zusätzlichen Verdampfer zwischen dem vorhandenen letzten Körper und dem barometrischen Kondensator einer Salzgewinnungsanlage auszulegen, zu bauen und zu betreiben, um einen möglichst großen Anteil dieser Abwärme zu nutzen.

Es erwies sich als möglich, etwa 40 % dieser Abwärme im Zusatzverdampfer zu nutzen, ohne die Kühlwassermenge zu erhöhen. Dadurch ergab sich eine 5 %ige Einsparung der für den Verdampfungsprozeß erforderlichen Gesamtenergie.

ÜBERSICHT ÜBER DIE FORSCHUNGS- UND ENTWICKLUNGSAKTIONEN DER EUROPÄISCHEN GEMEINSCHAFT ZUR NUTZUNG DER NIEDERTEMPERATURWÄRME

P.A. PILAVACHI,
Generaldirektion "Wissenschaft, Forschung und Entwicklung"
Kommission der Europäischen Gemeinschaften, Brüssel

Die Europäische Gemeinschaft hat seit 1979 im Rahmen von drei Energieeinsparungsprogrammen an Forschungsarbeiten auf dem Gebiet der Wärmerückgewinnung auf europäischer Ebene mitgewirkt.

Die Prozeßwärme, die von der Industrie in Abhängigkeit von der Temperatur benötigt wird, weist zwei Spitzen auf: zwischen 80°C und 200°C und zwischen 800°C und 1 400°C. Im gesamten Temperaturbereich werden große Mengen Abwärme freigesetzt, die, wenn sie zurückgewonnen und genutzt werden können, erhebliche Energieeinsparungen ermöglichen. Die Anwendungen der Wärmerückgewinnung für die niedrigere Spitze werden nachstehend beschrieben.

Der Teil, der die Wärmeanlagen betraf, konzentrierte sich auf Wärmetauscher und Wärmerohre, Kompressions- und Absorptionswärmepumpen, Wärmetransformatoren und ORC-Maschinen. Wärmerückgewinnungstechniken in der Textilindustrie, Metallverarbeitung, Papierindustrie, Speiseölherstellung sowie in Bäckereien, Wäschereien und anderen Betrieben zeigten, daß durch Rückgewinnung und Wiederverwendung von Wärme signifikante Energieeinsparungen möglich sind. Um das Problem einer schlechten Abstimmung zwischen Energieangebot und -nachfrage zu lösen, wurden auch Energiespeichertechnologien untersucht.

Weitere Aktivitäten können darauf ausgerichtet sein, eine Kapazität der Gemeinschaft bei der Fertigung kostenwirksamer Kompressions- und Absorptionswärmepumpen zu gewährleisten. Bei Wärmetauschern werden Verbesserungen bestehender Wärmetauscher sowie das Problem der Oberflächenverschmutzung als hohe Prioritäten angesehen, während neue Konzepte, einschließlich der Prozeßintensivierung, als wichtig angesehen werden.

PROZESSÜBERWACHUNG UND OPTIMIERUNG

B. KALITVENZEFF
Université de Liège, Lüttich

Das Fahren und die Steuerung industrieller Prozesse hat sich in den beiden letzten Jahrzehnten von zunächst unabhängigen und dann mehr oder weniger gekoppelten analogen Regelkreisen zur direkten Digitalsteuerung mit On-line-Computern, teilweise mit Entkoppelung der Regelkreise entwickelt. Auf dem Markt befinden sich adaptive und selbstregelnde Steuergeräte, die vermutlich mehr und mehr eingesetzt werden. Dank dem Einsatz von Mikrocomputern wird sich auch die hierarchische Steuerung allgemein durchsetzen.

Es gibt noch eine Menge zu tun, bevor On-line-Computer in der Lage sind, einen industriellen Prozeß tatsächlich zu steuern und noch mehr, um seinen Verlauf zu optimieren. Entsprechende Schritte werden ebenso wie für diskontinuierliche oder kontinuierliche Prozesse ergriffen; wir werden uns hauptsächlich mit dieser zweiten Art befassen.

Ein erster Schritt ist die Kenntnis des Ist-Standes des Prozesses mit Hilfe der in der Anlage erfaßten Informationen. Mehrere Fragen müssen beantwortet werden. Gibt es genügend Messungen, ist die Redundanz der Daten gut verteilt, welches sind die möglichen wirksamen Variablen, die gemessen werden müssen, wie sollen die Messungen im Hinblick auf die Deckung der Massen- und Energiebilanzen in Einklang gebracht werden ?

Ein unvermeidlicher Schritt ist die Analyse der Freiheitsgrade des Prozesses, die durch bestimmte Punkte "verbraucht" werden müssen, welche dem Steuerungssystem der Anlage vorgeschrieben werden: eine Antwort auf dieses Problem ist nicht Sache des Regeltechnikers, sondern des Konstrukteurs der Anlage oder des Betriebsleiters und erfordert ein eingehendes Verständnis des Prozesses.

Diese Analyse führt zu der Zuweisung der manipulierten Variablen zu den zu regelnden Variablen. Ein sehr wirksamer Weg, um dies zu erreichen, sind Sensitivitätsmatrizes. Dieses Leistungsfähige Werkzeug wird von der Industrie nicht genügend benutzt. Es kann durch Simulationen des stationären Zustands generiert werden. Die Zuweisungen können in Abhängigkeit von den Produktionsraten variieren: die Flexibilität der Prozeßsteuerung hat sich in den vergangenen Jahren ebenso wie die Flexibilität der Anlage als ein wichtiger Forschungsbereich erwiesen.

Alle diese Bedenken hängen eng mit der mathematischen Modellierung und der Simulierung des Verhaltens zusammen. Wir befassen uns hier selbstverständlich nicht mit Black-box-Modellen, sondern mit Wissensmodellen.

Wenn wir dann versuchen, den Betrieb der Anlage zu optimieren, ist die Simulierung des Verhaltens noch dringender notwendig, obwohl bei Verwendung von Black-box-Modellen einige nützliche Leitlinien gefunden werden können. Die Leistungswerte und -grenzen beider Arten von Modellen werden erörtert. Die Organisation von Prozeßdatenbanken ist so zu gestalten, daß der Betriebsleiter in der Lage ist, verschiedene Programme zur Beantwortung verschiedener Fragen hinsichtlich des Prozesses laufen zu lassen.

Mehrere Prozeßdatenbanken können zusammengeschlossen werden, um die notwendigen Dateien für den Koordinator der Energieversorgung des Standorts zu schaffen.

Erörtert wird auch eine dynamische Simulation, die als spezifisches Werkzeug zur Analyse komplexer Steuerungsprobleme - beispielsweise wenn mehrere Regelkreise einander stark beeinflussen - erforderlich ist.

Ein Vorschlag für ein koordiniertes prozeßrechner gestütztes System wird gegebenenfalls vorgelegt.

EIN NICHTKONVENTIONELLES PROJEKT ZUR RÜCKGEWINNUNG VON ENERGIE: ERZEUGUNG VON HOCHDRUCKDAMPF BEI DER OXYDATION VON PHOSPHOR

H.P. VAN HEEL
Hoechst, Holland - Vlissingen

Hoechst Holland N.V. nimmt Ende 1987 einen Hochdruckdampfkessel in Betrieb. Die bei der Oxydation von Elementarphosphor bei etwa 2 000°C freigesetzte Energie wird im Dampf mit einem Druck von 170 bar umgewandelt, der über ein mehr als 1 km langes Rohrleitungssystem in andere Produktionseinheiten eingespeist wird.

Dies ermöglichst die Aufheizung der Wärmeträger Marlotherm und Gilotherm, die bisher durch Erdgas aufgeheizt wurden, so daß etwa 14 Millionen m3 Erdgas eingespart werden.

Da eine Pufferspreicherung der zurückgewonnenen Energie nicht möglich ist, werden völlig getrennt betriebene Prozesse (Phosphorsäure, Dimethylterephtalate,Alkansulfonat) eng miteinander gekoppelt; hierfür sind elektronische Kommunikations- und Steuerungseinrichtungen sowie neue Managementfähigkeiten erforderlich.

Die in dem phosphorbeheizten Dampfkessel erzeugten "Gichtgase" (P205) werden im nachgeschalteten Turm durch Absorption in Wasser zum Hauptprodukt (Phosphorsäure).

Die Europäische Wirtschaftsgemeinschaft hat den innovativen Charakter dieses Demonstrationsvorhabens anerkannt und einen Zuschuß von 1 Mio ECU zugesagt.

EIN EXEMPLARISCHES PROJEKT: DAS OPTIMIERTE FAHREN EINER DAMPFKRACKANLAGE

V. KAIZER, HURSTEL
Technip, Paris

Die Erzeugung von Olefinen durch Dampfkracken von Kohlenwasserstoffen ist ein industrieller Schlüsselprozeß, da er die Grundlage der Kunststoff-, Faser- und chemischen Industrie ist.

Die Wirtschaftlichkeit der Operation hängt weitgehend vom Preis für Rohstoffe und Energie ab. Die Dampfkrackanlage erzeugt in der Regel mehrere Produkte mit hoher Wertschöpfung Ethylen, Propylbutadien und Benzol). Es ist daher von entscheidender Bedeutung, den Prozeßablauf zu optimieren, um der tatsächlichen Nachfrage des Marktes nach Ausgangsstoffen verschiedener Qualität gerecht zu werden.

Der Energieverbrauch beläuft sich auf etwa 2,5 MWh je tonne verarbeiteter Ausgangsstoff, d.h. auf 15-20% der technischen Kosten von Ethylen. Die Optimierung des Energieeinsatzes hat daher einen starken Einfluß auf die Wirtschaftlichkeit der Produktion.

Die große Zahl der bei der Steuerung dieses Prozesses beteiligten Variablen bedeutet, daß er sich besonders gut für eine Anwendung in Form eines computergestützten, fortschrittlichen Managementsystems eignet. Folgende Elemente müssen in Betracht gezogen werden:

- Erstellung eines computergestützten Modells der Anlage

- wirtschaftliche Optimierung der Produktion durch Verwendung des Modells eines fortschrittlichen Managementsystems

- Erfassung von Anlagendaten und Abstimmung zur Vorhersage optimierter Betriebsvariablen

- Vorhersage künftiger Operationen in Abhängigkeit vom Marktbedarf und der Verfügbarkeit von Ausgangsstoffen.

Die grundlegenden Elemente des fortschrittlichen Managementsystems werden vorgestellt und die erwarteten Ergebnisse ihrer Anwendung besprochen. Einige typische Aspekte dieser Art von Projekten werden erörtert.

MIKROPROZESSORSYSTEM MIT REGELSCHLEIFE ZUR VERBESSERUNG DER EFFIZIENTEN ENERGIENUTZUNG DER WINDERHITZER

A. SCIARETTA
Nuova Italsider, Taranto

Der Umbau des Hochofens 5 des Stahlwerks Taranto von Italsider hat zu bedeutenden Verbesserungen hinsichtlich des Energieverbrauchs, vor allem für die Winderhitzer und das Managementsystem geführt.

Am 24. Juni 1982 wurde der Hochofen erneut angefahren; zur Zeit erzeugt er mehr als 9 200 t Roheisen pro Tag.

Im Vergleich zu den Betriebsergebnissen der Ersten Lebensphase wurden bedeutende Fortschritte erzielt. Der Energieverbrauch lag in der ersten Lebensphase bei 560 Mcal/t Roheisen gegenüber 480 Mcal/t Roheisen in der derzeitigen Lebensphase.

Der energetische Wirkungsgrad, der in der ersten Lebensphase bei etwa 68-70% lag, wurde durch folgende Maßnahmen verbessert:

- Einbau eines neuen keramischen Brenners,
- Überprüfung der Verbrennung mit Hilfe der Anreicherung des Hochofengases für jeden Winderhitzer,
- Rückgewinnung der empfindlichen Wärme, der Rauchgase,
- automatisches Fahren mit Mikroprozessoren und Prozeßrechnern.

Dank dieser Art von Änderungen war eine Steigerung des Ausstoßes um etwa 5-7% möglich.

Während der zweiten Lebensphase werden neue Anwendungstechniken zur Steuerung und zum Fahren der Winderhitzer eingesetzt. Sie verwenden:

- Mikroprozessoren für die Direktsteuerung der Ventile und die Steuerung der Regelkreise,
- Prozeßrechner, die durch Datenfernverarbeitung mit den Mikroprozessoren verbunden sind, um den Prozeß zu überwachen und zu optimieren.

Es gibt zwei Interventionsebenen für die Anlage. Die erste, d.h. die Mikroprozessorebene, ermöglicht

- was die Anpassungen angeht, das Fahren der Aufheizphase jedes Winderhitzers(die für die Brennstoffanreicherung oder für die Anpassung der Verbrennungsparameter getrennt behandelt wird) unter Verringerung des Gesamtenergieverbrauche, einer optimalen Nutzung des Hochofengases und einer Verringerung des Verbrauchs an zusätzlichen Gasen mit höherer Verbrennungswärme (Naturgas, Koksofengas);
- was die Reihenfolge der Umstellungen angeht, das Fahren durch Überführen des Winderhitzers in den erforderlichen Zustand (Gas, Wind oder geschlossenes System) mit direkter Steuerung der Ventile und Intervention bei Betriebsstörung oder Alarm sowie Überführung der Anlage in einen sicheren Zustand.

Die zweite Interventionsebene, d.h. die der Prozeßrechner, ermöglicht eine weitere Verringerung des Gesamtenergieverbrauchs durch kontinuierliche und perfekte Überwachung der Flammentemperatur und durch die Beeinflussung der Dauer von Phasen und Wärmemengen durch mathematische Modelle.

NEUE TECHNIKEN FÜR DAS MANAGEMENT VON ENERGIEVERSORGUNGS-SYSTEMEN IN INDUSTRIEANLAGEN

G.B. ZORZOLI
Vorstand der ENEL, Roma

Neue Techniken für das Management von Energieversorgungssystemen in Industrieanlagen können in drei Kategorien unterteilt werden. Die erste befaßt sich mit Technologien, die entweder entwickelt wurden, um den Wirkungsgrad der Energieumwandlung oder die Benutzung ungewöhnlicher Brennstoffe wie Wärmepumpen in dem zuerst genannten und Wirbelschichtkessel in dem zuletzt genannten Fall zu verbessern. Sie umfaßt ferner bahnbrechende Technologien, deren industrielle Anwendung in naher zukunft erwartet wird (ein typischer Fall sind Brennstoffzellen). Die zweite Kategorie befaß sich mit fortschrittlichen Lösungen der Wärmekraftkopplung: von ORC-Maschinen, die sich für Wärmequellen mit geringer Enthalpie eignen, bis zu Energiekaskaden und integrierten Energiesystemen für Industriebezirke, die entweder durch konventionelle Brennstoffe oder durch vor Ort anfallende Abfälle und Energiequellen versorgt werden. Die dritte Kategorie befaßt sich mit der Entwicklung von Hard- und Software, die sich für die Optimierung und Steuerung von Prozessen in Industrieanlagen eignen und von speziellen Sensoren, die u.a. genauere Informationen über die Flammentemperatur in einem Brenner liefern sollen. Die mögliche Rolle der in den drei Kategorien enthaltenen Technologien wird sowohl getrennt als auch durch Prüfung ihrer Synergismen unter besonderer Berücksichtigung des Einflusses der Technologien der Kategorien 1 und 3 auf die Einführung spezifischer Lösungen für die Wärmekraftkopplung in der Industrie bewertet.

ANWENDUNGEN DER SOFTWARE SECI MANAGER ZUR OPTIMIERUNG VON ENERGIESYSTEMEN UND KONTINUIERLICHEN INDUSTRIEPROZESSEN

M. COEYTAUX
Serete, Paris

Die technische Komplexität von Energiesystemen in großen Industrieanlagen und die Vielzahl der Verträge über die Energieversorgung bieten Möglichkeiten für die Optimierung von Systemen, deren Abwicklung für Konstrukteure und Betreiber noch immer mit Schwierigkeiten verbunden ist. Vor kurzem entwickelte Werkzeuge für die nichtlinerare Modellierung und Optimierung wie die SECI Manager-Software sind nunmehr im Einsatz und helfen, Fortschritte auf diesem Gebiet zu erzielen. Eine gleichzeitige Betrachtung von Energiequellen und Anlagenprozessen sollte schon bald zur Gesamtoptimierung eines vollständigen Industriestandorts führen.

WÄRME-KRAFT-KOPPLUNG DURCH VERBRENNUNG VON RAFFINERIEGASEN

A. KALYVAS
Motor Oil Hellas, Athen

Die Raffinerie von Motor Oil (Hellas) in Korinth ist sowohl hinsichtlich der verwendeten Prozesse und Techniken als auch hinsichtlich des Spektrums und der Qualität der erzeugten Produkte eine komplexe Raffinerie. Die Leitung des Unternehmens beschloß 1983 ein Kraftwerk zu erstellen um die Energie der bis dahin "abgefackelten" überschüssigen Rauchgase einzusparen, die durch die häufigen Stromsperren im Netz der öffentlichen Elektrizitätsversorgung verursachten Verluste auf ein Mindestmaß zu beschränken und den hohen Elektrizitätsbedarf der Raffinerie zu decken. Das Kraftwerk nutzt die zuvor abgefackelten Gase für ein Gemisch mit den Brenngasen verschiedener Produktionseinheiten der Raffinerie.

Das Kraftwerk besteht aus :

1) zwei Turbinen und zwei Wechselstromgeneratoren zu je 11,5 MW und einer Temperatur von 35°C bei Druck auf Meereshöhe,

2) einem Brenngasverdichtungssystem und

3) einer Abwärmerückgewinnungs- und Dampferzeugungsanlage, die sowohl Hoch- als ach Niederdruckdampf erzeugt.

Die Anlage ist seit November 1984 in Betrieb. Auf der Ebene der Raffinerie wurden Energieeinsparungen von 9 193 Mio toe/Jahr erreicht; die nutzbaren Energieeinsparungen der Raffinerie beliefen sich auf 36 923 Mio toe/Jahr.

NUTZUNG VON SOFTWARE-SYSTEMEN ZUR OPTIMIERUNG DER WÄRME-KRAFT-KOPPLUNG

J. SPRINGELL
I C I, Billingham

Das Kraftwerk der ICI in Wilton on Teesside ist ein typisches Beispiel für viele industrielle Kombikraftwerke, die Elektrizität und mehrere Arten von Dampf erzeugen, um den fluktuierenden Energiebedarf eines Werkes zu decken. Die traditionellen Betriebsstrategien zur Bewältigung dieser Veränderungen waren zu stark vereinfacht und haben die Betriebskosten nur in geringem Maße oder überhaupt nicht berücksichtigt. Dieses Papier beschreibt HAMBLE, eine Anwendung fortgeschrittener, computergestützer Optimierungstechniken, die es dem Kraftwerk ermöglicht, den erforderlichen Bedarf zu minimalen Kosten zu decken. Das System berücksichtigt geänderte wirtschaftliche Grunddaten, interne Einzelheiten, Änderungen der Leistungskennwerte und betriebliche Sachzwänge.

Die Funktion von HAMBLE kann ausgeweitet werden, um eine wirksame Koordinierung der Energienachfrage und des Energieverbrauchs an einem Standort zu ermöglichen. Gesamt- und Grenzkosten werden ermittelt, und alternative Betriebsstrategien können rasch beurteilt werden.

Diese Fallstudie beschreibt die Rolle von HAMBLE bei der Planung und Durchführung der energiewirtschaftlichen Politik am Standort Wilton.

DIE TÄTIGKEIT DER KOMMISSION IM BEREICH DER FINANZINSTRUMENTE

H. CARRE und W. FABER
Kommission der Europäischen Gemeinschaften
Generaldirektion Wirtschaft und Finanzen, Brüssel

1. Thema

Dieses Thema ist für die Kommission neu. Es ist notwendig, in Europa ein System und Mechanismen zu förden, die nicht nur darauf ausgerichtet sind, europäischen Unternehmen finanzielle Produkte und Dienstleistungen zur Verfügung zu stellen, die sie für ihre Gründung, Entwicklung und Zusammenarbeit benötigen. Diese Mechanismen sollten es ferner gestatten, neue Finanzierungsmöglichkeiten zu erschließen, um die Realisierung von Vorhaben mit einer europäischen Dimension zu erleichtern.

2. Konzept

Im derseitigen Kontext ist es wünschenswert, die Beteiligung der öffentlichen Hand zu reduzieren und zugleich die verfügbaren Privatmittel zu mobilisieren. Es ist daher notwendig, als Katalysator für die Mobilisierung von Privatmitteln aufzutreten und den Markt dabei zu unterstützen, geeignete Mechanismen zu entwickeln, die darauf abgestellt sind, finanzielle Aktionen oder Vorhaben zu fördern, denen nach Ansicht der Gemeinschaft ein besonderes Interesse zukommt.

Auf Gemeinschaftsebene soll die Tätigkeit im Bereich der Finanzinstrumente das im Überfluß auf dem Markt verfügbare Kapital mobilisieren. Sie soll nicht an die Stelle des Marktes treten, sondern ihn dazu anhalten, geeignete Finanzierungsverfahren anzubieten. Die Gemeinschaft hofft, die Rolle eines Multiplikators für die Aufbringung der Gelder zu spielen die für die Förderung von Investitionen in der Gemeinschaft erforderlich sind.

3. Anwendung

Investitionen zur Verbesserung der effizienten Energienutzung kommt in der Gemeinschaft eine hohe Priorität zu, und die Kommission beabsichtigt, ihre Finanzierung zu erleichtern. Die hierfür erforderlichen Mechanismen werden zur Zeit untersucht. Sie können das Spektrum von der einfachen Bereitstellung von Zuschüssen bis zu ausgeklügelteren Systeme zum Abbau von Investitionshemmnissen, wie z.B. einen Garantiefonds umfassen.

DIE FINANZIERUNG DURCH DRITTE IN DER EUROPÄISCHEN GEMEINSCHAFT

D. FEE
Kommission der Europäischen Gemeinschaften
Generaldirektion Energie, Brüssel

- Gründe für die geringen Investitionen im Bereich der effizienten Energienutzung in der EG;

- Funktionsweise der Finanzierung durch Dritte;

- Potential der Finanzierung durch Dritte in der EG;

- Rolle der Kommission bei der Unterstützung der weiterverbreiteten Anwendung der Finanzierung durch Dritte.

EINE NEUE FINANZIERUNGSQUELLE FÜR ENERGIEEINSPARINVESTITIONEN

Dr. H. JUNKER
Bayerische Landesbank Girozentrale, München

Energie ist Lebensgrundlage für uns alle. Da wir nur über eine begrenzte Menge nicht erneuerbarer Energieresourcen verfügen, tritt das Ziel, mit Energie sparsam zu wirtschaften, zunehmend in den Vordergrund. Die in der EG zusammengeschlossenen Staaten streben einen weiteren 20%-igen Minderverbrauch bis 1995 an und bieten dafür vielfältige Fördermöglichkeiten.

Ein neues, unkonventionelles Finanzierungskonzept bietet Third Party Financing, ein fremdes Unternehmen führt die Investitionen durch und wird dafür aus den Kosteneinsparungen bezahlt.
Die Einführung des Konzeptes stösst u.a. auf folgende Schwierigkeiten:

- Energieeinsparinvestitionen erscheinen häufig wegen langer Amortisationszeiten als unrentabel;

- im Gebäudesektor wurden bereits Energieeinsparungen in nicht unerheblichem Masse durchgeführt, das verbleibende Einsparpotential wird vom Investor oft zurückhaltend beurteilt;

- in Finanzierungsformen wie Third Party Financing investieren relativ wenig Kapitalgeber wegen der schwer überschaubaren Risiken.

Für die Banken - die strengen Auflagen unterliegen - stellen sich zwei Gruppen von Kreditfällen dar. Falls als Kreditnehmer Träger öffentlicher Einrichtungen auftreten, wird das zu tragende Risiko minimiert. Bei privaten Unternehmen handelt es sich dagegen meist um kleinere bis mittlere Unternehmen, deren Bonität oft über die Dauer der gesamten Kreditlaufzeit nicht abschliessend beurteilt werden kann.

In letzteren Fällen wird über die Möglichkeit der Risikominimierung - z.B. über Bürgschaften der öffentlichen Hand - nachgedacht.

FINANZIERUNGSQUELLEN FÜR INVESTITIONEN ZUR VERBESSERUNG DER EFFIZIENTEN ENERGIENUTZUNG - AUS DER SICHT EINES INDUSTRIELLEN

D. KALYVAS
Motor Oil Hellas, Athen

Solche Investitionen sind häufig von großer Bedeutung im Hinblick auf das erforderliche Kapital sowie die Fähigkeit der Unternehmen zur Selbstfinanzierung. Sie werfen aber gleichzeitig nur durchschnittliche Gewinne ab, obwohl die Ergebnisse für die effiziente Energienutzung von großer Bedeutung sind. Es ist daher klar, daß zur Realisierung solcher Investitionen eine langfristige Finanzierung erforderlich ist.

Obwohl staatliche Beihilfen häufig gewährt werden, sind Bankkredite notwendig. Eine Kreditaufnahme ist in den meisten Fällen unmöglich oder sehr kostspielig, da die Unternehmen den privaten und öffentlichen Kredit instituten ausreichende Garantien bieten müssen.

Die Studie und Realisierung eines solchen Instruments auf Gemeinschaftsebene, d.h. eines Garantiesystems, das die Überwindung dieser schwierigkeiten ermöglicht, wird zweifellos zu einer Erhöhung der Investitionen im Bereich der effizienten Energienutzung führen.

RESUMES EN LANGUE FRANCAISE

Introduction

Intégration des procédés

Nouvelles techniques de récupération de la chaleur basse température

Installations industrielles - Contrôle et optimisation des procédés

Gestion énergétique des utilités

Table ronde sur le financement des investissements d'efficacité énergétique

PRINCIPES FONDAMENTAUX DE LA CONSERVATION ENERGETIQUE

H. SCHAEFER,
Institut für Energiewirtschaft und
Kraftwerkstechnik, München

Dans chaque cas, quel que soit l'ordre de grandeur, la rationalisation de la consommation énergétique dans les procédés industriels doit être envisagée selon cinq axes :

- élimination de la consommation inutile,
- réduction de la consommation d'énergie utile,
- amélioration du rendement des machines et des installations,
- récupération de l'énergie, lorsque ce recyclage est techniquement et économiquement envisageable,
- développement de l'utilisation des énergies renouvelables.

L'analyse de la situation énergétique du site concerné constitue une condition préalable incontournable à toute approche visant à une utilisation plus rationnelle de l'énergie. Pour constituer une base valable, une telle analyse doit reposer sur des mesures réalisées en service, qu'il s'agisse de machines ou d'installations individuelles, ou de l'ensemble d'une unité industrielle. Dans de nombreux cas, il faudra introduire de nouvelles technologies mais, souvent, il suffira de coordonner et d'intégrer les connaissances disponibles et des techniques éprouvées.
Dans l'ensemble, les mesures d'économie peuvent concerner l'ingénierie énergétique et de la production, ainsi que les paramètres d'organisation et de fonctionnement; normalement, les mesures envisagées concerneront l'ensemble de ces domaines. Le personnel responsable de ces secteurs devra donc travailler en étroite coopération pour mettre ces mesures en oeuvre et assurer leur continuité.

Etant donné les différences importantes dans le rapport coût/bénéfice offert par les diverses options, chaque cas d'optimisation de l'utilisation énergétique doit être évalué spécifiquement.

Toutefois, une attention particulière doit être accordée aux effets cumulatifs ou neutralisants de mesures différentes prises dans les mêmes domaines. Il faut en outre tenir compte des interactions complexes entre les modifications apportées au niveau de la production et/ou de la productivité, d'une part, et la demande énergétique, de l'autre.

Enfin, il faut savoir que dans de nombreux cas l'énergie est remplacée par le capital et que, comme dans un même temps la technologie devient de plus en plus complexe, les besoins de personnel hautement qualifié et formé spécifiquement constitueront un facteur important.

CONSERVATION DE L'ENERGIE DANS LA FABRICATION DES VILEBREQUINS : UN EXEMPLE D'ANALYSE INTEGREE BASEE SUR DES MESURES DETAILLEES

M. RUDOLPH, Institut für Energiewirtschaft und Kraftwerkstechnik, München

Pour évaluer le potentiel d'économie énergétique dans un cas particulier, il convient de mesurer et d'analyser toutes les données relatives à la consommation d'énergie, aux performances des installations, etc.

Dans le cas qui nous occupe, une étude a été réalisée sur deux chaînes d'usinage de vilebrequins. Deux précisions importantes s'imposent :

- les pièces non usinées sont forgées pour une chaîne, et coulées pour l'autre,
- l'une des chaînes a peu d'articulations, tandis que l'autre est une chaîne de transfert.

Pour diverses raisons, les chiffres relatifs à la consommation énergétique par vilebrequin diffèrent de plus de 20 %. Ces différences importantes sont dues au poids des pièces finies ainsi qu'à la quantité de matériau découpée, aux capacités de production des deux chaînes et aux rapports entre le temps d'usinage et d'inactivité. Il faut également envisager les différentes consommations énergétiques des mécanismes de transport. Un certain nombre d'unités ne tire en effet pas pleinement parti des caractéristiques du moteur d'entraînement.

A certains stades de l'usinage, différentes techniques de découpage du métal sont appliquées dans un même objectif.

Du point de vue énergétique, divers moyens permettent de réduire considérablement la consommation : mise hors circuit des unités inactives, optimisation des caractéristiques des moteurs d'entraînement, techniques de pointe de régulation de la vitesse du moteur et choix optimal des méthodes de découpage. Cependant, une comparaison entre les coûts et les avantages de ces mesures montre que le potentiel réel de conservation est nettement moins important.

INTEGRATION DES PROCEDES A L'AIDE DE LA TECHNOLOGIE 'PINCH'

LINNHOFF & A. EASTWOOD
University of Manchester
Institute of Science and Technology, Manchester

Au cours de ces cinq dernières années, il est apparu que l'intégration des procédés permet d'obtenir des améliorations importantes au niveau de la conception des installations industrielles. Grâce à cette intégration, la construction et l'exploitation des installations sont moins coûteuses et plus aisées.

Un outil a permis de réaliser ces progrès dans la conception : la technologie 'Pinch' qui, si elle s'appuie sur la thermodynamique fondamentale, est cependant facile à comprendre et à utiliser.

Cette communication décrit brièvement la base scientifique de la technologie 'Pinch' et montre comment elle est appliquée.

Nous présentons deux études de cas qui montrent clairement les avantages de l'intégration des procédés dans le contexte de la récupération de chaleur. La première étude met en lumière le fait que la récupération de chaleur doit être envisagée dans le contexte global d'une installation (et, de préférence, de l'ensemble d'un complexe industriel) si l'on veut réaliser les économies adéquates. Des économies énergétiques qui semblent évidentes et justifiées lorsqu'on les envisage isolément peuvent en effet entraîner la perte de certains potentiels et réduire les possibilités d'économie dans d'autres secteurs.

La seconde étude de cas montre comment les connaissances acquises par le truchement de la technologie 'Pinch' peuvent être utilisées pour modifier un procédé afin de réaliser des économies énergétiques.

Enfin, nous donnons une indication quant aux économies types qui peuvent être réalisées par l'utilisation de cette technologie dans le cadre de nouveaux projets et de rénovations, en nous référant à des études qui ont été effectuées au cours des cinq dernières années.

INTEGRATION DE PROCEDES DANS UNE RAFFINERIE DE BENZOL

R.L. BARDSLEY
Staveley Chemicals Ltd, Chesterfield

Staveley Chemicals exploite six installations de traitement chimique continu près de Chesterfield, dans le Derbyshire.

L'installation BTC a fait l'objet d'études en matière de conservation énergétique. Le benzène, le toluène et le cyclohexane sont produits dans cette installation en raffinant du benzol brut par un procédé catalytique à haute température.

Une étude de six mois sur l'application de la technologie pinch a donné les résultats suivants avec une consommation énergétique de base de 1 000 unités :

	BTC	Unité lythol	récupération des gaz de fumée
Utilisation actuelle	1 000	543	543
Cible énergétique minimum	860	421	299
Cible énergétique pratique			323

Il semblait possible d'obtenir en pratique une diminution de 40 % de la consommation énergétique en incluant la récupération de la chaleur des gaz de fumée.

Les coûts et la rentabilité ont été évalués pour diverses options; on a en outre dressé un bilan de la répartition énergétique sur l'ensemble du site. Ces travaux ont débouché sur une recommandation de changement du système à vapeur et d'installation de deux turbines.

Projet 2		
Six échangeurs de chaleur : lythol	supérieur à 100 000 UKL	environ un an
Projet 3		
Deux turbines, production combinée électricité-chaleur	Nettement supérieur au projet 2	Moins de quatre ans

La compagnie prévoit une économie annuelle de 750 000 UKL réalisée grâce à la mise en oeuvre de ces projets.

La technologie pinch s'est avérée vitale pour la révision de la conception d'un réseau complexe d'échange de chaleur afin de réduire la consommation énergétique au minimum. On a pu dégager quelques options claires pour lesquelles les coûts et les avantages potentiels ont pu être évalués.

INTEGRATION DES PROCESSUS DANS UNE BRASSERIE

R. MARSH
Tetley Walker Ltd, Warrington

Cette communication montre comment les consultants en matière d'énergie, Linnhoff March Ltd., commandités par l'office britannique pour l'éfficacité énergétique ont appliqué la technologie 'Pinch' aux procédés consommateurs d'énergie dans une brasserie britannique typique afin de réduire la consommation et les coûts qui y sont associés.

Après une présentation de la société concernée et de ses produits, nous décrivons comment la brasserie, qui faisait déjà une utilisation très efficace de l'énergie, a pu accroître son rendement énergétique en appliquant les résultats d'une étude selon la technologie 'Pinch' réalisée par Linnhoff March Ltd.

Illustrée par des diapositives, la communication décrit en détail les applications pratiques des méthodes d'économie appliquées pour conserver la vapeur, le gaz naturel et le fuel. Nous montrons comment trois démarches possibles ont été identifiées et, élément important, quelles étaient exactement les cibles énergétiques de chaque démarche pour l'ensemble de la brasserie.

La présentation se termine par un résumé des leçons tirées de l'application de la technologie 'Pinch' aux processus de brassage et envisage comment cette technologie peut être appliquée à d'autres procédés industriels dont les coûts énergétiques sont élevées.

TECHNOLOGIE DE LA RECOMPRESSION MECANIQUE DE VAPEUR ET DES POMPES A CHALEUR

A. GLUCKMAN
March Consulting Group, Windsor

Au cours de ces dix dernières années, plusieurs centaines de systèmes industriels de pompes à chaleur ont été installés en Europe.

Les performances de ces systèmes sont très variables. Certaines installations ont donné d'excellents résultats pratiques et économiques, tandis que d'autres, moins satisfaisantes, présentent une faible rentabilité et des problèmes de fiabilité.

Nombre de leçons intéressantes peuvent être tirées de ces installations existantes.

Cette communication présente un bref aperçu des pompes à chaleur industrielles existantes. Ces informations permettent de dégager des lignes directrices pour une conception adéquate des pompes à chaleur. La communication met plus particulièrement en lumière certains aspects de la conception du système et des composants qui doivent toujours être envisagés aux premiers stades d'un projet de ce type. Les marchés de la pompe à chaleur et les applications en Europe sont également envisagés.

IMPACT DES NOUVELLES TECHNOLOGIES SUR LA CONCEPTION FUTURE DES ECHANGEURS DE CHALEUR

D. REAY
David Reay & Associates, Whitley Bay

La technologie de la récupération de chaleur est très largement représentée dans la plupart des grands programmes industriels de conservation de l'énergie, et la pénétration dans les processus des équipements d'échange de chaleur est extrêmement importante. Certaines entraves techniques et économiques font cependant encore obstacle à la généralisation des échangeurs de chaleur. On peut citer par exemple les problèmes liés à l'encrassement et à la corrosion; on est aussi parfois tenté d'adopter des méthodes de récupération plus complexes, comme les machines à cycle Rankine organique, ce qui débouche souvent sur des projets trop ambitieux offrant des avantages économiques plus que relatifs.

Les perspectives de récupération rentable de chaleur à basse température s'améliorent grâce à toute une série de développements, et notamment aux matériaux et concepts nouveaux pour les échangeurs de chaleur. Un certain nombre d'aides périphériques, et notamment l'amélioration de la conception et le recours à l'intelligence artificielle, permettent aux utilisateurs d'opérer une sélection appropriée parmi les équipements les plus perfectionnés.

L'intensification des procédés qui concerne normalement les installations chimiques compactes, aura un rôle important à jouer à l'avenir en réduisant la dimension et les coûts des échangeurs de chaleur. Ce concept sera décrit dans la communication qui aborde également les problèmes de récupération de chaleur à basse température et envisage l'impact possible des nouvelles technologies sur la conception future des échangeurs de chaleur.

RECUPERATION D'ENERGIE PAR RECOMPRESSION MECANIQUE DE LA VAPEUR DES HYDROCARBURES

J.P. LIVERNET
Rhône Poulenc, Chalampé

L'objectif de cette réalisation est de récupérer la vapeur organique au sommet d'une colonne de distillation pour la compresser à une pression suffisante et la condenser dans les rebouilleurs de la même colonne au lieu d'utiliser la vapeur à 6 bars du réseau de l'usine.

Cette installation est intégrée dans une unité de production de cyclohexanol-1 qui est en activité depuis 1972 aux installations Rhône Poulenc de Chalampé (68, France).

L'unité comporte essentiellement un compresseur, des échangeurs de chaleur et la tuyauterie; vu les risques de corrosion, l'ensemble de l'équipement, y compris le compresseur, est réalisé en acier inoxydable.

Le bilan énergétique de la colonne montre que l'économie de vapeur s'élève à 32 T/h et que la consommation d'électricité par le compresseur est de 3 400 kW/h.

Les études de conception ont été réalisées par le département central d'ingénierie de Rhône Poulenc. Un modèle réduit complet (1/33) a été utilisé pour les études détaillées et pour former le personnel d'exploitation.

Le coût total s'élevait à 41 millions de FF en janvier 1985. En raison de ses aspects innovateurs, ce projet a bénéficié d'aides octroyées par la Commission des Communautés européennes et par l'agence française pour la maîtrise de l'énergie.

En dépit de l'importance des investissements, liée à la nature des fluides qui sont inflammables et corrosifs, l'opération est un succès.

ECHANGEURS DE CHALEUR EN PLASTIQUE

J. HUYGHE
G.R.E.Th., C.E.N.
Grenoble

La récupération de chaleur à basse température, procédé qui semble souvent très simple par rapport au procédé à haute température, rencontre cependant certaines difficultés. Il y a tout d'abord le coût des échangeurs de chaleur et. dans de nombreux cas, les problèmes d'encrassement et de corrosion de la surface d'échange.

Grâce au développement des matériaux plastiques, on peut maintenant envisager d'utiliser ces matériaux pour les applications industrielles à basse température (moins de 150°C). afin de réaliser des échangeurs de chaleur qui permettent de résoudre ces problèmes.

Différents types d'échangeurs de chaleur en plastique ont donc été développés en Europe depuis quelques années. Grâce à leur conception originale, certains échangeurs de chaleur réduisent au minimum les inconvénients dus à la faible conductivité thermique du matériau. Ces échangeurs présentent, par rapport aux échangeurs métalliques, les avantages suivants :côut généralement moindre, légèreté, meilleure résistance à la corrosion et, dans certains cas, encrassement réduit en raison du faible pouvoir d'adhésion des produits d'encrassement sur les matériaux plastiques.

Cette communication passe en revue les différents types d'échangeurs de chaleur en plastique développés au cours de ces dernières années. La plupart d'entre eux sont maintenant commercialisés pour des températures en service pouvant aller justu'à 100 ou 150°C.

RECOMPRESSION DE LA VAPEUR DANS UNE BRASSERIE

E. NOLTING
M.A.N. Technologies GmbH, Munich

1. Introduction

Le premier compresseur hélicoïde de vapeur entraîné par un moteur au gaz naturel a été installé à la brasserie Dortmunder Kronen en Allemagne. Les critères initiaux pour l'optimisation du procédés d'ébullition du moût portaient sur la réduction des émissions et les économies énergétiques.

2. Description du système de recompression de la vapeur

Un compresseur hélicoïde entraîné par un moteur au gaz naturel est utilisé pour compresser la vapeur. L'unité constituée du moteur et du compresseur a été installée sur une plate-forme au-dessus des deux chaudières à houblonner et équipée d'une protection acoustique. L'unité est complètement isolée contre les vibrations et il n'y a donc pas de transfert notable de bruits de structure dans le bâtiment. La chaudière Thermostar supplémentaire est suspendue derrière les deux chaudières à houblonner.

3. Fonction du système de recompression de la vapeur

Le système de recompression de la vapeur travaille alternativement avec la chaudière 1 et la chaudière 2. Le taux d'évaporation est préréglé par l'ajustement de la vitesse du moteur à gaz. La vapeur s'élevant de la chaudière à 1 bar et 100°C est aspirée par le compresseur hélicoïde et compressée à 1,3 - 1.6 bars. La vapeur se condense dans la chaudière Thermostar et fournit la chaleur au moût qui est ensuite renvoyé dans la chaudière à houblonner.

4. Expériences au niveau de l'exploitation

Le potentiel de réduction des émissions et d'économies d'énergie a été confirmé en service permanent. Le système fonctionne depuis plus de 5 000 heures.

VALORISATION DE LA VAPEUR RESIDUELLE PROVENANT DE L'EVAPORATION DE LA SAUMURE

P.E. BUNGE
Akzo Zout Chemie, Hengelo

Des trains d'évaporateurs sont utilisées pour produire du sel sec à partir de la saumure. La saumure est réchauffée dans le premier évaporateur par passage de vapeur à basse pression dans l'échangeur de chaleur qui lui est associé. La vapeur de la saumure en ébullition dans le premier évaporateur est utilisée pour chauffer la saumure dans l'évaporateur suivant où la pression est inférieure. Cette opération se répète tout au long du train d'évaporateurs et la vapeur du dernier évaporateur est condensée par un système de refroidissement par eau. Pour la production du sel, on utilise normalement de quatre à six évaporateurs.

Si le prix de l'énergie augmente, l'installation d'un évaporateur supplémentaire en amont du premier peut être rentable.

Toutefois, lorsque l'usine de production de sel est reliée à une installation de production combinée électricité-chaleur, l'avantage offert par la consommation réduite de vapeur est à peu près perdu parce que l'évaporateur supplémentaire exigera de la vapeur à haute pression et que l'énergie produite sera donc moins importante. De plus, la température du premier évaporateur doit être limitée pour éviter la corrosion.

La vapeur provenant du dernier évaporateur a une température de condensation de 43°C seulement mais, comme le volume disponible est important, la quantité de chaleur est considérable. Cette chaleur est normalement considérée comme de la chaleur résiduelle en raison de sa basse température. Notre tâche concernait la conception, la construction et l'exploitation d'un évaporateur supplémentaire situé entre le dernier évaporateur existant et le condensateur barométrique d'une installation de production de sel afin de récupérer une quantité aussi importante que possible de cette chaleur résiduelle.

On a constaté qu'il était possible d'utiliser environ 40 % de la chaleur résiduelle dans l'évaporateur supplémentaire sans accroître la quantité d'eau de refroidissement nécessaire. Cette opération a permis d'obtenir une économie de 5 % de l'énergie totale requise pour le procédés d'évaporation.

PANORAMA DES ACTIONS DE RECHERCHE ET DE DEVELOPPEMENT DE LA COMMUNAUTE EUROPEENNE EN MATIERE DE RECUPERATION DE CHALEUR A BASSE TEMPERATURE

P.A. PILAVACHI
Direction générale "Science, Recherche et Développement"
Commission des Communautées européennes, Bruxelles

La Communauté européenne est impliquée dans la recherche européenne en matière de récupération de chaleur depuis 1975, dans le cadre de trois programmes de conservation d'énergie.

La quantité de chaleur industrielle requise dans l'industrie en fonction de la température présente deux pics, l'un situé entre 80°C et 200°C, et l'autre, entre 800°C et 1 400°C. Sur toute la plage des températures, de grandes quantités de chaleur résiduelle sont évacuées qui, si elles pouvaient être récupérées et utilisées, permettraient de réaliser d'importantes économies d'énergie. Ce sont les applications de récupération de chaleur au niveau du pic inférieur qui seront décrites.

La section qui concerne l'équipement est centrée sur les échangeurs de chaleur et les canalisations, les pompes à chaleur à compression et absorption, les transformateurs de chaleur et les machines à cycle de Rankine organique. Des technologies de stockage d'énergie ont également été étudiées pour résoudre les problèmes entre l'offre et la demande. Les techniques de récupération de chaleur utilisées notamment dans les secteurs des textiles, du traitement des métaux, du papier, des huiles alimentaires, de la boulangerie et des lavoirs montrent que des économies considérables peuvent être réalisées grâce à la récupération et au recyclage de la chaleur.

A l'avenir, les activités pourront être orientées pour assurer que la Communauté dispose des capacités requises dans le domaine de la fabrication de pompes à chaleur industrielles à compression et absorption qui soient rentables. On considère que l'amélioration des échangeurs de chaleur existants et la solution des problèmes d'encrassement sont prioritaires; les nouveaux concepts, notamment d'intensification des procédés, sont également importants.

COMMANDE DE PROCEDES ET OPTIMISATION

B. KALITVENTZEFF
Université de Liège, Liège

La gestion et la commande des procédés industriels ont évolué au cours de ces deux dernières décennies, depuis les boucles d'asservissement analogiques, d'abord indépendantes puis de plus en plus souvent interconnectées, jusqu'aux commandes numériques directes avec ordinateurs travaillant en temps réel, parfois chargés du découplage des boucles. Les contrôleurs adaptatifs et auto-ajustables sont présents sur le marché et seront probablement mis en oeuvre de plus en plus fréquemment. Les commandes hiérarchiques vont également se généraliser grâce à l'utilisation des micro-ordinateurs.

Des progrès importants doivent encore être réalisés pour que les ordinateurs soient en mesure de contrôler réellement un procédé industriel et d'optimiser davantage encore son fonctionnement. On s'efforce de réaliser cet objectif, tant pour le traitement par lot que pour le traitement continu; c'est essentiellement ce second type de procédé que nous aborderons.

Dans un premier temps, il faut déterminer l'état du procédé suivant les informations relatives aux installations. Il faut répondre à plusieurs questions : le nombre de mesures effectuées est-il suffisant, la redondance des données est-elle bien répartie, quelles sont les variables efficaces à mesurer, comment réconcilier les mesures afin de satisfaire les bilans de masse et d'énergie ?

Etape inévitable, il faut analyser les degrés de liberté des procédés, ceux-ci qui doivent être "consommés" par les points imposés au système de commande de l'installation : c'est au concepteur ou au gestionnaire des installations qu'il incombe de résoudre ce problème et non pas à l'ingénieur responsable des commandes, et cette opération exige une connaissance parfaite du procédé.

Cette analyse débouche sur l'affectation des variables traitées aux variables à contrôler. Les matrices de sensibilité permettent de réaliser cette opération de façon très efficace. L'industrie ne tire cependant pas suffisamment parti des possibilités offertes par cet outil puissant. Il peut être produit par des simulations de régime permanent. Les affectations peuvent varier en fonction des rythmes de production : depuis quelques années, la souplesse des commandes de procédés s'est avérée, comme la souplesse d'opérabilité des installations, un domaine de recherche important.

Tous ces éléments sont étroitement liés à la modélisation mathématique et à la simulation du comportement. Il est évident qu'il ne s'agit pas ici de modèles "boîte noire" mais de modèles de connaissance.

Si nous nous efforçons d'optimiser le fonctionnement des installations, la simulation des caractéristiques est davantage encore nécessaire, bien que certaines lignes directrices puissent être déterminées à l'aide des modèles "boîte noire". Nous envisageons les performances et les limites des deux types de modèles. L'organisation des bases de données du procédé doit être telle que le gestionnaire des installations sera en mesure de faire tourner des logiciels différents pour répondre à diverses questions relatives à un même procédé.

Plusieurs bases de données sur le procédé peuvent être fusionnées pour créer les fichiers de données nécessaires aux coordinateurs de la gestion des utilitaires du site.

Nous envisageons également la simulation dynamique requise pour l'analyse des problèmes complexes de commande lorsque, par exemple, des boucles d'asservissement ont entre elles un jeu complexe d'interactions.

Enfin, nous présentons une proposition de système coordonné pour la gestion de procédé assistée par ordinateur.

UN PROJET ORIGINAL DE RECUPERATION DE CHALEUR : PRODUCTION DE VAPEUR A HAUTE PRESSION DANS UN PROCEDE D'OXYDATION DU PHOSPHORE

H.P. VAN HEEL
Hoechst Holland, Vlissingen

Hoechst Holland N.V. a mis en service à la fin de 1987 une chaudière à vapeur à haute pression. L'énergie produite durant l'oxydation du phosphore élémentaire à environ 2 000°C est convertie en vapeur à 170 bars qui est envoyée à d'autres unités de traitement via un système de canalisation de plus d'un kilomètre.

Ceci permettra de chauffer les caloporteurs Marlotherm et Gilotherm (huiles), chauffage qui jusqu'à présent se faisait au gaz naturel. Cette opération permettra d'économiser environ 14 millions de m3 de gaz naturel.

Comme il n'est pas possible de stocker l'énergie récupérée, des procédés tout à fait différents (acide phosphorique, diméthyltéréphtalate, alkane-sulphorate) seront couplés; cette opération exige un équipement électronique de communication et de commande, ainsi que de nouvelles compétences de gestion.

Les "gaz de cheminées" (P2O5) produits dans la chaudière constituent le produit principal (acide phosphorique) dans la tour en aval, par absorption dans l'eau.

La Communauté économique européenne a reconnu le caractère novateur de ce projet de démonstration et a promis un financement d'un million d'Ecus.

UN PROJET EXEMPLAIRE : LA GESTION OPTIMISEE D'UN VAPOCRAQUEUR

V. KAIZER, HURSTEL
Technip, Paris

La production d'oléfines par le vapocraquage d'hydrocarbures constitue un procédé industriel clef parce qu'il fournit la base nécessaire aux industries des plastiques, des fibres et des produits chimiques.

Les économies d'exploitation dépendent très largement du prix des matières premières et de l'énergie. Le vapocraqueur produit, typiquement, plusieurs produits de haute valeur ajoutée (éthylène, propylène-butadien et benzène); il est donc essentiel d'optimiser le traitement afin de répondre à la demande du marché avec des matières premières de qualités différentes.

La consommation énergétique s'élève à quelque 2,5 MWh par tonne de matières traitées, soit 15 à 20 % du coût technique de l'éthylène. L'optimisation de l'apport énergétique a donc une influence considérable sur le coût de production.

En raison du grand nombre de variables impliquées dans la commande de ce procédé, ce dernier est particulièrement adapté à la mise en oeuvre d'un système de pointe de gestion informatisée (AMS). Les éléments suivants doivent être envisagés :

établissement d'un modèle informatisé des installations ;

optimisation économique de la production grâce au modèle AMS;

collecte des données relatives aux installations qui sont mises en concordance avec les prévisions des variables de fonctionnement optimisées;

prévisions des opérations futures en fonction des besoins du marché et de la disponibilité des matières premières.

Nous présentons les éléments essentiels du système de gestion AMS et envisageons les résultats attendus de son application. Certains aspects caractéristiques de ce type de projet sont discutés.

SYSTEME A MICROPROCESSEURS ET BOUCLES DE REGULATION NUMERIQUES POUR ACCROITRE LES ECONOMIES D'ENERGIE DANS LES FOURS

A. SCIARETTA
Nuova Italsider, Taranto

La reconstruction du haut-fourneau 5 des installations sidérurgiques d'Italsider à Tarente a introduit des améliorations importantes au niveau de la consommation énergétique, particulièrement pour les fours et le système de gestion.

Le 24 juin 1982, le haut-fourneau a été remis en service et il produit à l'heure actuelle plus de 9 200 tonnes de fonte par jour. Par rapport à la situation antérieure, cette rénovation a permis d'obtenir des résultats opérationnels importants.

La consommation énergétique dans les premières installations s'élevait à 560 000 Kcal/tonne de fonte produite alors qu'elle s'élève aujourd'hui à 420 000 Kcal/tonne.

L'amélioration du rendement énergétique, qui était d'environ 68-70 % avec les premières installations, est due :

- à l'installation d'un nouveau brûleur céramique;
- au contrôle de la combustion avec enrichissement du gaz de haut-fourneau pour chaque four;
- à la récupération de la chaleur des fumées;
- à la gestion automatique avec microprocesseurs et traitement informatique.

Ces modifications permettent d'obtenir une augmentation de rendement d'environ 5 à 7 %.

Les nouvelles technologies sont appliquées de la façon suivante au contrôle et à la gestion des installations de la seconde génération :

- microprocesseurs pour la commande directe des vannes et le contrôle des boucles de régulation;
- ordinateur connecté par télétraitement aux microprocesseurs afin de superviser et d'optimiser le procédé.

Il y a deux niveaux d'intervention; le premier, qui concerne les microprocesseurs, permet :

- au plan des réglages, la gestion de la phase de chauffage de chaque four (traitée séparément soit pour l'enrichissement du combustible, soit pour le réglage des paramètres de combustion) se fait par réduction de la consommation énergétique totale, optimisation de l'utilisation des gaz de haut-fourneau et réduction de la consommation des gaz supplémentaires à plus haute température de combustion (gaz naturel, gaz des fours à coke).

- Au plan des séquences ou des inversions, la gestion se fait en amenant le four à l'état voulu (gaz, vent ou fumée), en commandant directement les vannes et en intervenant en cas de défaillance ou d'alarme et en plaçant les installations en mode sécurité.

Le second niveau d'intervention, qui concerne l'ordinateur, permet de réduire davantage encore la consommation énergétique totale grâce au contrôle continu et optimal de la température de la flamme et à une intervention, par le biais de modèles mathématiques, dans la durée des phases et au niveau des quantités de chaleur concernées.

NOUVELLES TECHNIQUES POUR LA GESTION DES UTILITES DANS LES INSTALLATIONS INDUSTRIELLES

G.B. ZORZOLI
Board of Directors, ENEL, Rome

Les nouvelles techniques de gestion des utilités en installations industrielles peuvent être réparties en trois classes. La première concerne les technologies mises au point pour améliorer soit le rendement de la conversion énergétique, soit l'utilisation de systèmes "exotiques", comme les pompes à chaleur dans le premier cas et, dans le dernier cas, les chaudières à lit fluidisé. Elle inclut également des technologies de pointe dont l'application industrielle est prévue dans un proche avenir (les piles à combustible en sont un exemple type). La seconde classe concerne les solutions nouvelles pour la production combinée d'électricité et de chaleur : depuis les cycles Rankine organiques convenant aux sources de chaleur basse enthalpie, jusqu'aux systèmes énergétiques en cascade et aux systèmes intégrés pour les zones industrielles alimentés par des combustibles conventionnels, par la chaleur résiduelle ou par des sources énergétiques locales.

La troisième classe concerne la mise au point de matériel et de logiciel permettant d'optimiser et de contrôler des procédés en installations industrielles, ainsi que de capteurs spéciaux fournissant, par exemple, des informations plus exactes sur la température de la flamme d'un brûleur. Le rôle envisageable pour les technologies utilisées dans les trois classes est évalué individuellement, d'une part, et en étudiant leur synergie, de l'autre, en accordant une attention toute particulière à l'influence des technologies des classes 1 et 3 sur la pénétration dans le secteur industriel de solutions spécifiques pour la production combinée de chaleur et d'électricité.

APPLICATION DU LOGICIEL DE GESTION SECI A L'OPTIMISATION DES SYSTEMES ENERGETIQUES

M. COEYTAUX
Serete, Paris

La complexité technique des systèmes énergétiques dans les installations industrielles à grande échelle et la diversité des contrats d'approvisionnement en énergie offrent des perspectives d'optimisation des systèmes que les concepteurs et les opérateurs considèrent encore comme difficiles à gérer. Des outils non linéaires de modélisation et d'optimisation mis au point récemment, comme le logiciel SECI-MANAGER sont maintenant en exploitation et permettent de réaliser des progrès dans ce domaine. Un examen simultané des sources énergétiques et des procédés industriels devrait déboucher très rapidement sur l'optimisation de l'ensemble d'un site industriel.

PRODUCTION COMBINEE D'ELECTRICITE ET DE CHALEUR PAR COMBUSTION DES GAZ DE RAFFINERIE

A. KALYVAS
Motor Oil Hellas, Athènes

La raffinerie Motor Oil (Grèce) à Corinthe est une raffinerie complexe tant en ce qui concerne les divers procédés et techniques utilisés qu'au niveau de la gamme et de la qualité des produits fabriqués. La gestion de la société a décidé en 1983 d'installer une centrale électrique afin de récupérer l'énergie des gaz résiduels de combustion brûlés à la torche et de minimiser les pertes entraînées par les coupures fréquentes du réseau public, tout en répondant aux besoins accrus d'électricité de la raffinerie. Cette centrale électrique utilise les gaz jadis brûlés à la torche afin de les mélanger aux gaz de combustion provenant des diverses unités de production de la raffinerie.

La centrale comporte (1) deux turbines et deux alternateurs de 11,5 MW chacun à 35°C à la pression au niveau de la mer, (2) un système de compression des gaz et (3) une unité de récupération de la chaleur résiduelle et de production de vapeur produisant de la vapeur à haute et à basse pression. Les installations fonctionnent depuis novembre 1984. Les économies énergétiques réalisées au niveau de la raffinerie s'élèvent à 9 139 MTEP par an et les économies énergétiques utilisables représentent 36 923 MTEP par an.

UTILISATION DE SYSTEMES LOGICIELS POUR OPTIMISER LA PRODUCTION COMBINEE D'ELECTRICITE ET DE CHALEUR

J. SPRINGELL
I C I, Billingham

La centrale ICI à Wilton, Teesside, constitue un exemple type de nombreuses centrales industrielles de production combinée fournissant l'électricité et plusieurs types de vapeur pour répondre à la demande énergétique fluctuante d'une usine. Traditionellement, les stratégies d'exploitation pour régler ces problèmes de fluctuation sont simplistes et tiennent peu ou ne tiennent pas compte des coûts d'exploitation. Cette communication décrit le système HAMBLE, une application des techniques de pointe d'optimisation assistées par ordinateur qui permet à la centrale de répondre à la demande pour un coût minimum. Le système tient compte des paramètres économiques variables, des détails internes ainsi que des variations des performances et des contraintes d'exploitation.

La fonction de HAMBLE peut être étendue pour permettre la coordination efficace de l'approvisionnement et de la consommation énergétique sur l'ensemble d'un site. Les coûts totaux et marginaux sont communiqués et des stratégies opérationnelles de remplacement peuvent être rapidement évaluées.
Cette étude de cas décrit le rôle de HAMBLE dans la planification et la mise en oeuvre de la politique de gestion énergétique au site de Wilton.

LES ACTIVITES DE LA COMMISSION DANS LE DOMAINE DE L'INGENIERIE FINANCIERE

H. CARRE, W. FABER
Commission des Communautés européennes
Direction générale Affaires économiques et financières, Bruxelles

1. Thème

Ce thème est nouveau pour la Commission. Il convient de promouvoir en Europe un système et des mécanismes qui permettent non seulement de fournir aux sociétés européennes les produits et services financiers dont elles ont besoin pour la création, le développement et la coopération. Ces mécanismes doivent permettre d'ouvrir de nouvelles perspectives financières afin de faciliter la réalisation des projets de dimension européenne.

2. Démarche

Le contexte actuel est marqué par un souci de réduire l'engagement budgétaire du secteur public en mobilisant, dans un même temps, les fonds privés disponibles. Il est donc nécessaire d'agir en tant que catalyseur pour mobiliser ces fonds privés et en assistant le marché pour développer des mécanismes appropriés destinés à promouvoir les actions financières ou les projets auxquels la Communauté attache un intérêt tout particulier.

Au niveau communautaire, les activités d'ingénierie financière visent à mobiliser la masse considérable de capitaux disponibles sur le marché. Elle n'a pas l'intention de se substituer au marché mais d'inciter le dernier à offrir des formes adéquates de financement. La Commission entend jouer un rôle multiplicateur ou de levier pour mobiliser les fonds requis afin de stimuler les investissements dans la Communauté.

3. Application

Les investissements en matière de rendement énergétique sont hautement prioritaires dans la Communauté et la Commission a l'intention de faciliter leur financement. Les mécanismes permettant de concrétiser cette démarche sont actuellement à l'étude. Ils peuvent être multiples, de la simple attribution d'aides jusqu'aux programmes plus élaborés visant à éliminer des obstacles à l'investissement, comme par exemple un fonds de garantie.

D. FEE
Commission des Communautés européennes
Direction générale Energie, Bruxelles

D. FEE a abordé la question du financement par des tiers dans les Communautés européennes.

- Raisons du faible niveau des investissements consacrés au rendement énergétique dans la CEE;
- fonctionnement du financement par des tiers;
- potentiel de ce type de financement dans la CEE;
- rôle de la Commission dans la généralisation du système de financement par des tiers.

UNE NOUVELLE SOURCE DE FINANCEMENT POUR LES INVESTISSEMENTS DANS LES ECONOMIES D'ENERGIE

H. JUNKER
Bayerische Landesbank Girozentrale München

L'énergie est vitale pour nous tous. Etant donné que nous ne disposons que d'une quantité limitée de sources d'énergie non renouvelable, la nécessité d'user de l'énergie avec économie est de plus en plus pressante. Les Etats membres de la CE s'efforcent d'obtenir une nouvelle réduction de 20 % de la consommation d'ici à 1995 et proposent à cette fin diverses possibilités.

Le financement par des tiers (Third Party Financing) offre une nouvelle méthode de financement originale : une entreprise tierce réalise les investissements et elle est payée au moyen des économies réalisées sur les coûts.

L'introduction de cette méthode se heurte aux difficultés suivantes :

- les investissements dans les économies d'énergie se révèlent souvent non rentables en raison des longs délais d'amortissement;
- des économies d'énergie non négligeables ont déjà été réalisées dans le secteur du bâtiment, le potentiel d'économie d'énergie restant suscite souvent des réticences chez l'investisseur;
- les investisseurs optent relativement peu pour les formes de financement telles que le Third Party Financing en raison de la difficulté d'appréciation des risques.

Pour les banques qui sont soumises à de sévères contraintes, il y a deux catégories en matière de crédit : lorsque l'emprunteur est un organisme public, le risque à supporter est réduit au minimum. Dans le secteur privé, il s'agit en revanche le plus souvent de petites et moyennes entreprises dont la solvabilité pendant toute la durée du crédit ne peut être appréciée de façon certaine.

Dans ce dernier cas, on peut envisager la possibilité de réduire le risque au minimum, par exemple au moyen de cautionnements du secteur public.

FINANCEMENT DES INVESTISSEMENTS D'EFFICACITE ENERGETIQUE - POINT DE VUE D'UN INDUSTRIEL

D. KALYVAS
Motor Oil Hellas, Athènes

Bien souvent les investissements de ce type exigent des capitaux considérables, par rapport aux capacités d'autofinancement des entreprises, mais n'offrent dans un même temps qu'une rentabilité moyenne, même si les résultats en matière de rendement énergétique sont importants. Il est donc évident que le financement à long terme est nécessaire pour réaliser ces investissements.

Même si l'Etat accorde souvent une aide, les emprunts bancaires deviennent une nécessité. Cette démarche est dans la plupart des cas impossible ou extrêmement coûteuse parce que les entreprises doivent fournir des garanties suffisantes pour satisfaire aux exigences des banques commerciales ou institutionnelles européennes.

L'étude et la réalisation au niveau communautaire d'un tel outil - un système de garantie - permettant de surmonter ces difficultés entraîneront de toute évidence une augmentation des investissements en matière de rendement énergétique.

LIST OF PARTICIPANTS

Alexandre, J.
Ministère de la Région Wallonne
DAETN - Inspection Générale de l'Energie
Ave du Prince de Liège 7
B - 5100 Jambes

Alonso Ruiz, J.J.
CADEM, S.A.
C/San Vicente, 8
Edificio Albia, I-planta 14
E - 48001 Bilbao

Anargyrou, G.
Eichhorster Weg 12
D - 1000 Berlin 26

Andersen, H.M.
Centec Technology Consultants
7 Falkoner Allee
P.O. Box 95
DK - 2000 Frederiksberg

Angerer, G.
Fraunhofer-Institut für Systemtechnik
und Innovationsforschung
Breslauer Str. 48
D - 7500 Karlsruhe 1

Arend, P.
SOBEMAP
Place du Champ de Mars 5/Boite 40
B - 1050 Brussels

Asmann, H.
Hoechst AG
D - 6230 Frankfurt/Main

Aspichueta Larruscain, J.J.
Altos Hornos de Viscaya
4 Carmen 2
E - 48901 Baracaldo (Vizcaya)

Aussenhofer, W.
Energieberatung
Philosophenweg 8
D - 3500 Kassel

Awerbuch, M.
Consulting Office
Hafenstr. 18
D - 1000 Berlin 33

Awerbuch, N.
S.A.R.L. Supratherm
3 Passage des Entrepreneurs
F - 75015 Paris

Baker, C.H.
ICI (Chemicals & Polymers Group)
The Heath Runcorn
P.O. Box 14
GB - Cheshire WA 7 4QG

Bardsley, R.L.
Staveley Chemicals
NR Chesterfield
GB - Derbyshire S 43 2 PB

Barendregt, S.
TECHNIP
Cedex 23
F - 92090 Paris la Défense

Barone, F.
ARBED S.A.
B.P. 142
L - 4008 Esch-sur-Alzette

Baumann, R.
Ing.büro f.optimierte Energieanwendung
Zeitblomstr. 47/1
D - 7900 Ulm/Donau

Beck, P.
Umweltbundesamt
Bismarkplatz 1
D - 1000 Berlin 33

Van den **Berg**, H.
Industrielle Energie Analysen
SCK/CEN
Boeretang 200
B - 2400 Mol

Biffin, M.
University College
Mechanical Engineering
Cardiff - Newport Road CF2 ITA
GB - South Glam, Wales

Bokelmann, H.
GEA GmbH
Rensingstr. 9
D - 4630 Bochum

Bork, P.-W.
BEWAG
Motzstr. 89
D - 1000 Berlin 30

Born, P.
Energieberatung GmbH
Richard-Wagner Str. 41
D - 4300 Essen 1

Brandt, P.
IG Metall Verwaltungsstelle Wattingen
Grosse Weilstr. 8
D - 4320 Wattingen

Brath, P.
COWIconsult
Teknikerbyen 45
DK - 2830 Virum

Briganti, G.
ENEA CACACCIA
St. Anguillarese
I - 00060 Roma

Brönd, E.
NESA A/S
Vaerebrovej 109
DK - 2880 Bagsvaerd

Broschk, J.
Rheinisch-Westfäl. Elektrizitätswerk AG
Abt. Anwendungstechnik
Kruppstr. 5
D - 4300 Essen 1

Bunge, P.E.
Akzo Zout Chemie Nederland bv
Postbus 25
NL - 7550 GC Hengelo Ov.

Buscaglione, A.
UNAPACE
Via Paraguay 2
I - 00198 Roma RM

Van **Buuren**, J.E.
Hoogovens Groep BV
P.O. Box 10000
NL - 1970 CA Tjmuiden

Carstens, Th.
Bonnenberg & Drescher
Elssholzstr. 22
D - 1000 Berlin 30

Casals, J.M.
Colomer Investigacion y Desarrollo, S.A.
P.O. Box 15
St. Francesc 1
E - 08500 VIC

Castanheira de Carvalho, J.J.
Cimpor-Cimentos de Portugal, E.P.
Rua Alexandre Herculano 35
P - 200 Lisboa

Cittadini, M.
LEGLERTEX S.P.A.
Via San Clemente 53
I - 24036 Ponte S. Pietro (BG)

Clausius, E.
BP Deutschland AG
Samberger Str. 4
D - 8070 Ingolstadt

Coeytaux, M.
SERETE
86 Rue Requault
F - 75640 Paris Cedex 13

Corini, G.
P.O. Box 125
I - Rieti

Coursimault, O.
Electricité de France (EDF) SEPAC
2 Rue Louis Murat
F - 75384 Paris Cedex 08

De **Crevoisier**, Ph.
Pont-A-Mousson S.A.
B.P. 45
F - 54703 Pont-A-Mousson S.A.

Custers, E.
Directorate General for Energy
Commission of the European Communities
Rue de la Loi 200
B - 1049 Brussels

Daclin, M.
S.A.R.L. Capital 1.012.5.00 F
58 rue Roger Salengro, Peripole 116
F - 94126 Fontenay sous Bois Cedex

Dadda, E.
Carlo Gavazzi Impianti
C.P. 47
I - 20013 Magenta (MI)

Davis, M.
Directorate-General for Energy
Commission of the European Communities
Rue de la Loi 200
B - 1049 Brussels

Dielmann, K.P.
Vereinigte Glaswerke GmbH
Viktoriaallee 3-5
D - 5100 Aachen

Dorino, E.
FIAT AUTO S.P.A.
Via la Manta 24
I - 10137 Torino

Dusserre, P.
Centre de Recherches Textiles de Mulhouse
185 Rue de l' Illberg
F - 68093 Mulhouse Cedex

Dyon, J.
Rhone Poulenc Industrialisation
24 Avenue Jean Jaures
F - 69151 Décines Cedex

Ericsson, E.
Schwedische Botschaft
Referat für Technik u.Wissenschaft
Heussallee 2-10
D - 5300 Bonn 1

Esbensen, T.
Planum International
Möllegade 54-56
DK - 6400 Sönderborg

Faber, W.
DG for Financial and Economic Affairs
Commission of the European Communities
B - 1043 Brussels

Fee, D.
Directorate General for Energy
Commission of the European Communities
Rue de la Loi 200
B - 1049 Brussels

Fedele Dell' Oste, E.
Egidio
ENEA CASACCIA
St. Anguillarese
I - 00060 Roma

Franck, P.A.
Chalmers University of Technology
S - 41296 Gothenburg

Galland, D.
Saint-Gobain STEV
Les Miroir Cedex 27
F - 92096 Paris la Defense

Garain, J.-P.
Ciments d'Obourg S.A.
B - 7048 Obourg

Gaurier, L.
Gaz de France
Direction des Etudes et
Techniques Nouvelles
361 Ave du Président Wilson
F - 93211 La Plaine St. Denis Cedex

Gazay Unibaso, L.A.
Altos Hornos de Vizcaya S.A.
C/Carmen 2
E - 48901 Baracaldo Vizcaya

Gbur, P.
Siemens
Mannheimer Str. 32
D - 1000 Berlin 31

van **Gemert**, P.H.
Centrum voor Energiebesparing en
Schone Technologie
Oude Delft 180
NL - 2611 HH Delft

Georgakopoulos, D.
EKO ABEE
Athens Tower
2 Messagion Ave
GR - 11527 Athens

Gilbert, Ph.
CDF Energie
Tour Albert 1°
65 Ave de Colmar
F - 92507 Rueil Malmaison Cedex

Gillet, A.
ALATAL S.A.
153 Ave Fond Roy
B - 1120 Bruxelles

Gioli, G.
APRICA STUDI Srl
Via Lamarmora 230
I - 25123 Brescia

Glatzel, W.-D.
Umweltbundesamt
Bismarkplatz 1
D - 1000 Berlin 33

Gluckmann, Th.R.
University of Manchester
Institute of Science and Technology
P.O. Box 88
GB - Manchester M 60 1QD

Goeminne, G.
Directorate General for Energy
Commission of the European Communities
Rue de la Loi 200
B - 1049 Brussels

Gomez Pascual, C.
CADEM, S.A.
C/San Vicente 8
Edificio Albia I - planta 14
E - 48001 Bilbao

Goublomme, B.
PRODIRA S.A.
98-100 Chaussée de Vilvorde
B - 1120 Bruxelles

Gourlin, J.P.
Centre de Recherche ELF-Solazi
B.P. 22
F - 69360 Saint Symphonien d'Ozon

Graas, André
SOLLAC
F- 57191 Florange Cedex

Gram, N.O.
Federation of Danish Industries
H.C. Andersens Boulevard 18
DK - 1596 Copenhagen V

Grehier, A.
Institut Francais du Pétrole
B.P. 311
F - 92506 Rueil Malmaison Cedex

Grossin, R.
Bertin & Cie
B.P. 3
F - 78373 Plaisir Cedex

Günther, H.
BASF Aktiengesellschaft
RED - C 100
D - 6700 Ludwigshafen

Hadjichristou, X.
Electromatic Constructions Ltd.
P.O. Box 3522
CY - Nicosia, Cyprus

Hannemann, H.
Raiffeisenverband
Raiffeisenstr. 1
D - 2300 Kiel 1

Hassemer, H.
VEGLA Vereinigte Glaswerke GmbH
Concordiaplatz 3
D - 5000 Köln 90

Van **Heel**, H.P.
Hoechst Holland N.V.
Werk Vlissingen
Postfach 65
NL - 4380 AB Vlissingen

Herbst, D.
Dr. Walter Herbst AG
Haynauer Str. 47
D - 1000 Berlin 46

Herrmann, J.
GEA - Wärmeaustauscher Happel GmbH & Co.
Niemetzstr. 41-45
D - 1000 Berlin 44

Hirs, G.G.
Comprimo Consulting Services B.V.
Postfach 4129
NL - 1009 AC Amsterdam

Hoogendoorn, A.
State University of Utrecht
Croeseskraat 77a
NL - 3522 AD Utrecht

Hummel, B.
Optimierte Energieanwendung
Zeitblomstr. 34
D - 7900 Ulm

Huyghe, J.
GRETH C.E.N.G. 85 X
F - 38041 Grenoble Cedex

Istikolglou, I.
EKO-Chemicals CO.A.E.
P.O. Box 10044
GR - 541 10 Thessaloniki

Jahn, A.
Deisterpfad 24
D - 1000 Berlin 37

Jandeaux, G.
ESSO Energie
B.P. 52
F - 76330 N.D. De Gravenchon

Johansson, M.
DK - Teknik
Gladsaxe Mollevej 15
DK - 2860 Soborg

Jones, H.
Cambridge University
Department of Engineering
Trumpington Street
GB - Cambridge CB4 3DB

Johnson, D.E.
BSC General Steels, Scunthorpe Works
PO box 1, Scunthorpe
GB - South Humberside, DN 16 BP.

Jongema, P.,
AKZO Zont Chemie Nederland
P.O. Box 25
NL - 7550 GC Hengelo CO1

Junker, H.-J.
Bayer Landesbank Girozentrale
Brienner Str. 20
D - 8000 München 2

Kaizer, V.
TECHNIP
Cedex 23
F - 92090 Paris la Défense

Kalatzis, A.
Ing.Büro Alexander Kalatzis
Erenstrole 7
GR - 26224 Patpas

Kalitventzeff, B.
Université de Liege
Institut de Chimie - Sart Tilman - B 6
B - 4000 Liege

Kalyvas, Ph.
Motor Oil (Hellas) Corinth Refineries S.A.
2 Karageorgi Servias Str.
GR - 10033 Athens

Kampet, T.
InnoTec GmbH Berlin
Kurfürstendamm 180
D - 1000 Berlin 15

Keet, F.A.M.
Nederlandse Philips Bedrijven B.V.
Cederlaan 4, gebouw OBE
NL - 5616 SC Eindhoven

Kellner, K.
IEA International Energy Agency
OECD
2 Rue André-Pascal (16e)
F - 75775 Paris Cedex 16

Khazaeli, F.
Bierstädter Str. 29c
D - 6200 Wiesbaden

Kilde, N.A.
Riso National Laboratory
P.O. Box 49
DK - 4000 Roskilde

Kleefkens, O.
Ministry of Economic Affairs
Department of Energy
P.O. Box 20102
NL - 2500 EC Den Haag

Klemm, W.
Benckiser-Knapsack GmbH
Dr. Albert Reimann Str. 2
D - 6802 Ladenburg/Neckar

Klemt, R.
Der Senator für Wirtschaft und Arbeit
Martin-Luther-Str. 105
D - 1000 Berlin 62

Kohrs, H.
Schlingmeier Quarzsand GmbH & Co. KG.
Ackerstr. 8-10
D - 3301 Schwülper 4

Kohrs, R.
Schlingmeier Quarzsand GmbH & Co. KG
Ackerstr. 8-10
D - 3301 Schwülper 4

Kopidiatis, E.
Bautechnische Anwendungen der
Umgebungsenergie
GR - Iraklion Kriti

Krause, H.
Hartmann & Braun
Vertrieb Minden Abt. VET
Schillerstr. 72
D - 4950 Minden

Kurka, H.-J.
3M Deutschland GmbH
Carl-Schurz-Str. 1
D - 4040 Neuss 1

Kyriakopoulos, L.
Athens Paper Mill S.A.
P.O. Box 3367
GR - 1020 Athens

Ledwon, E.
Verband Deutscher Maschinen
u. Anlagenbau e.V. VDMA
Lyoner Str. 18
D - 6000 Frankfurt/Main 71

Le Goff, P.
Institut National Polytechnique de Lorraine
Rue Grandville
F - 54042 Nancy Cedex

Linnhoff, B.
University of Manchester
Institute of Science and Technology
P.O. Box 88
GB - Manchester M60 1OD

Lintzen, R.
Bundeswirtschaftsministerium
Villemombler Str. 76
D - 5300 Bonn 1

Livernet, J.-P.
Societé Rhone-Poulenc Chimie
Usine de Chalampe
B.P. 267
F - 68055 Mulhouse Cedex

Llobell, R.
ELF France
Tour ELF (3.24-G-44)
F - 92078 Paris la Defense Cedex 45

Lopez Gisbert, B.
General Motors Espana, S.A.
Apartado de Correos 375
E - 50080 Zaragoza

Loureiro Campos, E.
Cimpor-Ciments de Portugal, E.P.
Rua Alexandre Herculano 35
P - 1200 Lisboa

Lucas-Herzfeld, S.
Gneisenaustr. 49
D - 1000 Berlin 61

Lupetti, C.
MONTEDIPE/Montediso
Petrochemical Company
Via Rosellini 15/17
P.O. Box 10779
I - 20110 Milano

Luxton, R.E. (SAM)
University of Adelaide
G.P.O. 498
AUS - 5001 Adelaide

Mafflioli, R.
ENICHEM ANIC
Piazza Boldrini 1
I - 20097 S. Donato Milanese

Malusardi, E.
SELM S.P.A. - SEE/TLI/ITE
Societa 'Energia Montedison S.P.A.
Via Taramelli 26
I - 20124 Milano

Maly, V.
Kernforschungsanlage Jülich GmbH
PBE P.O. Box 1913
D - 5170 Jülich

Maniatopoulos, C.S.
Directorate General for Energy
Commission of the European Communities
Rue de la Loi 200
B - 1049 Brussels

Manco, D.
Via Cicogna Mozzoni 6
I - 2061 Milano

Margull, E.
Wolff Walsrode AG
Postfach
D - 3030 Walsrode 1

Marsh, R.
Tetley-Walker Ltd.
The Brewery
GB - 312331 Warrington WA2 7NU

Mart, C.
Nederlandse Energie Ontwikkelings
Maatschappy BV Sittard/ NEOM
Postbus 17
NL - 6130 AA Sittard

Martin, G.
Rhone Poulenc Chimie de Base
25 Quai Paul Doumer
F - 92408 Courbevoie

Mechler, K.
Haus Bockdorf
D - 4152 Kempen 1

Melin, S.
Statens Vattenfall
S - 16287 Vällingby

Meraviglia, I.
Azienda dei Servizi Municipalizzati
del Comune di Brescia
Via Lamarmora 230
I - 25123 Brescia

Michel, A.
Athens Papermill S.A.
P.O. Box 3367
GR - 10210 Athens

Millet, B.
CEREN Centre d'Etudes et de Recherches
Economiques sur l'Energie
89 Rue de Miromesnil
F- 7500 8 Paris

Moerdijk, M.C.W.
Eco-Energy Engineering B.V.
Kleine Oord 87
NL - 6811 HZ Arnhem

Moisidis, K.
Phosphoric Fertilizers Industry
NEA Karvali
GR - 65110 Kavala

Morovic,
Fraunhofer-Institut für Systemtechnik
und Innovationsforschung (ISI)
Breslauer Str. 48
D - 7500 Karlsruhe 1

Mosser, F.
Industequip. S.A.
1 Rue du 22 Novembre
F - 67000 Strasbourg

Mülkens, W.
Bundesverband der Deutschen Industrie e.V.
Gustav-Heinemann-Ufer 84-88
D - 5000 Köln 51

Munk Nielsen, M.
Danish Ministry of Energy
Slotsholmsgade I
DK - 1216 Kobenhavn K.

Nieuwlaar, E.
State University of Utrecht
Croeseskraat 77a
NL - 3522 AD Utrecht

Nolting, E.
Man-Technologie
D - 8000 München

De **Oliveira**, R.
General Direction of Energy
Avenida da Republica 45-50
P - 1000 Lisboa

Ortenblad, W.
B & S Group
Aaboulevarden 22
Postfach 270
DK - 8100 Arkus C.

Paraskevas, G.
Hellenic Steel Co.
P.O. Box 10347
GR - 54110 Thessaloniki

Paules, B.
ATOCHEM - Usine de Gonfreville
B.P. 86
F - 76700 Harfleur

Petzold, D.
TU Berlin
Knesebeckstr. 1-2
D - 1000 Berlin 12

Pietravalle, G.
ITALSIDER S.P.A.
c/o Italsider - Via Appia
I - 74100 Taranto

Pilavachi, P.
Directorate-General for Science, Research and Development
Commission of the European Communities
200 Rue de la Loi
B - 1049 Brussels

Proiettei, G.
Telespazio S.P.A.
Via a. Bergamini 50
I - 00159 Roma

Raes, D.
GfE - Société Belge de Gestion d'Energie S.A.
Holstraat 61/A1
B - 8790 Waregem

Reay, Th.D.
David Reay & Associates
P.O. Box 25
Tyne & Wear
GB - Whitley Bay NE26 IQT

Römer, U.
ibek GmbH
Waterbergstr. 11
D - 2800 Bremen 21

Romagnoli, M.
ENICHEM ANIC
Piazza Boldrini 1
S. Donato Milanese
I - 20093 Milano

Romani, R.
ENEA - FARE - ENINT
C. P. 2400
I - 00100 Roma

Rudolph, M.
Technische Universität München
Energiewirtschaft und Kraftwerkstechnik
Arcisstr. 21
D - 8000 München

Ruiz Fernandez de Castro, J.M.
General Motors Espana, S.A.
Apartado de Correos 375
E - 50080 Zaragoza

Rütten-Freimuth, R.
MAN Technologie GmbH
Dachauer Str. 667
D - 8000 München 50

Sabbioni, F.
TESI
Via de Rolandi 11
I - 20156 Milano

Sakellarios, J.
Energie- und Industrieberatung
Psychiko Kokoni Str. 23
GR - 15452 Athens

Sandamuris, M.
Mpikou 4 Plastira pl.
GR - Athens

Sarris, N.H.
Phosphoric Fertilizers Industry Ltd.
Sygrou Av. 97
GR - 11745 Athens

Schaeffer, H.
Technische Universität München
Energiewirtschaft und Kraftwerkstechnik
Arcisstr. 21
D - 8000 München 2

Schiller
BASF AG
RE/C - Z 13, Wöhlerstr. 24b
D - 6700 Ludwigshafen

Schmidt, H.
Presse
Postfach 370225
D - 1000 Berlin 37

Schneider, H.
Ingenieurbüro Schneider
Siemensstr. 62
D - 7016 Gerlingen

Schnieder, H.
Daimler Benz AG
Daimler Str. 123
D - 1000 Berlin 48

Schön, E.
Ruhrgas AG
Huttropstr. 60
D - 4300 Essen 1

Schön, M.
Fraunhofer-Institut für Systemtechnik
und Innovationsforschung (ISI)
Breslauer Str. 48
D - 7500 Karlsruhe 1

Sciaretta, A.
ITALSIDER S.P.A.
Via Appia
I - 74100 Taranto

Silberhorn, O.
Ingenieurbüro Oswald Silberhorn
Ludwigstr. 26
Postfach 111847
D - 8900 Augsburg

Simnacher, R.
Optimierte Energieanwendung
Ing.-Büro
Ulmer Str. 3
D - 7916 Niersingen

Sirchis, J.
Directorate General for Energy
Commission of the European Communities
Rue de la Loi 200
B - 1049 Brussels

Sorensen, Stig Niemi
M.Sc.M.E.
Federation of Danish Industries
H.C. Andersens Boulevard 18
DK - 1596 Copenhagen V

Sotiriadis, Th.
MANOS S.A.
9 Pireos Str.
GR - 18346 Moshaton

Spoto, Maurizio
ENICHEM S.P.A.
Piazza Boldrini
I - 20097 San Donato Milanese

Springell, J.
Imperial Chemicals Industries
P.O. Box 86/2700 AB
NL - Zoetermeer

Staller, K.G.
Kinetics Technology International B.V.
P.O. Box 86/2700 AB
NL - Zoetermeer

Stone, G. J.
The Electricity Council
30 Millbank
GB - London SWIP 4RD

Straeter, D.
Ruhrgas AG
Huttropstr. 60
D - 4300 Essen 1

Theodoridis, M.
Phosphoric Fertilizers Industry Ltd
Nea Karvali
GR - 65110 Kavala

Tringali-Casanuova, L.
ANIE - Associazione Nazionale Industrie
Via Algardi 2
I - 20148 Milano KI

Trioullier, D.
Companie Générale de CHAUFFE
118-120 Rue de Rivoli
F - 75001 Paris

Turrini, Elisabeth
Provinzstr. 57
D - 1000 Berlin 51

Turner, G.
Senator für Wissenschaft und Forschung
Bretschneiderstr. 5
D - 1000 Berlin 19

Tzannetakis, P.
Motor Oil Hellas
2 Kabageorgi Servias Str.
GR - Athens

Unterwurzacher, E.
OECD-IEA
2 Rue André-Pascal
F - 75755 Paros Cedex 10

Urie, M.
Syndicat Intercommunal
A Vocation Multiple de Marie-Galante
B.P. 48
F - 97112 Grand-Bourg Marie-Galante

Vacchelli, G.
Industrielle Berater
Via Vigentina 21
I - 20122 Milano

Veenman, D.
Vy-Group Int. B.V.
P.O. Box 207
D - 7640 AA Wierden

Welcker, U.
Papiertechnische Stiftung PTS
Heßstr. 130a
D - 8000 München 40

Wendler, J.
HERBST Service GmbH
Haynauer Str. 47
D - 1000 Berlin 46

Wenzel, W.
BEWAG Abt. Fernwärme TFP
Potsdamer Str. 58
D - 1000 Berlin 30

Weinert, K.-H.
Interatom GmbH
Friedrich-Ebert-Str.
D - 5060 Bergisch Gladbach 1

Werner, L.
Gewerkschaft Auguste Victoria
Postfach 1180
D - 4370 Marl

Wille, W.
GAL Fernwärmeschiene Saar West
Besitzgesellschaft mbH Beteiligungsg.schaft
Bismarckstr. 11
D - 6620 Völklingen

Zito, U.
Directorate General for Energy
Commission of the European Communities
Rue de la Loi 200
B - 1049 Brussels

Zorzoli, G.B.
ENEL
Via G.B. Martini 3
I - 00198 Roma

INDEX OF AUTHORS

BARDSLEY, R.L., 52
BARENDREGT, S., 149
BRIGANTI, G., 5
BUNGE, P.F., 107

CARRE, H., 200
COEYTAUX, M., 178

DAVIS, M., 219

EASTWOOD, A., 37

FABER, W., 200
FEE, D.A., 204
FOSTER, D., 192

GLUCKMAN, R., 62

HURSTEL, X., 149
HUYGHE, J., 88

JUNKER, J., 208

KAISER, V., 149
KALITVENTZEFF, B., 130
KALYVAS, A., 185
KALYVAS, P., 211

LINNHOFF, B., 37
LIVERNET, J.P., 83

MANIATOPOULOS, C.S., 8
MARSH, R., 56

NOLTING, E., 98

PILAVACHI, P.A., 116

REAY, D.A. 73
RUDOLPH, M., 28

SCHAEFER, H., 14
SCIARRETTA, A., 159
SPRINGELL, J., 192

TURNER, G., 2

VAN HEEL, H.P., 141

ZORZOLI, G.B., 170